Handbook of Contract Management in Construction

Ali D. Haidar

Handbook of Contract Management in Construction

Ali D. Haidar
Dar Al Riyadh
Riyadh, Saudi Arabia

ISBN 978-3-030-72267-8 ISBN 978-3-030-72265-4 (eBook)
https://doi.org/10.1007/978-3-030-72265-4

This Springer imprint is published by the registered company Springer Nature Switzerland AG
The registered company address is: Gewerbestrasse 11, 6330 Cham, Switzerland

Preface

Contract management is the process that enables both parties to a contract to meet their obligations in order to deliver the objectives required from the contract. It also involves creating a good working relationship between the parties.

Contract management continues before and throughout the life of a contract and involves managing proactively the anticipation and the future needs of the job required as well as reacting to situations that arise.

One of the aims of contract management is to obtain the services as agreed in the contract and achieve value for money. This means optimising the efficiency, effectiveness and economy of the service and the relationship described in the contract; balancing costs against risks; and actively managing the relationships in a construction project between the employer and the service providers, for example, the contractor, the consultants and others including the nominated subcontractors and suppliers.

The terms and conditions of the contract in construction include the general conditions, the special conditions, specifications, bill of quantities (boq or BOQ), means to measure the items executed, price adjustment procedures, variation/change control procedures, suspension and termination clauses and all the other formal mechanisms that enable the employer to have control over the project.

The terms ‘contract administration’ and ‘contract management’ are often used interchangeably. Although they share many similarities, they are technically different phases of the contract lifecycle. In the construction industry, however, the same team of professionals may be responsible for both facets of the contracting process; which, it is likely, is one of the reasons we tend to conflate these terms (Corey 2015).

At its simplest, a contract is a document describing a relationship between two parties, their mutual agreement and who carries the risks if things do not turn out as planned. Contract management is about managing the relationship and the risks in the contract between the parties, to ensure that all parties get the result they originally wanted. All matters such as cash flow, revenue, obligations, responsibilities and liabilities flow from this objective (Cook 2014).

For ease and practicality, contract management is the topic that is addressed in this book; however, the reader might find some items can be addressed under contract administration.

Book Structure

In this book, Chapter 1 provides an introduction to contract management. Chapter 2 entails a comprehensive understanding of construction law, the main principles of formation of contract, and the doctrines a contract manager must understand in the event of a breach of contract.

Chapter 3 addresses the relationship between the parties to a project, and the framework of a properly structured contract within the general contract principles. Lump sum, cost plus, time and materials and guaranteed maximum price contracts, which define the basic project delivery system, are also reviewed.

Chapter 4 will review the different types of standard forms of contract usually used in the construction industry. These include the main standard forms of contract generated from the USA,[1] the UK[2] and internationally.[3]

Chapter 5 will look at the essential elements of contract drafting and the main conditions found in a typical contract, to give the reader a broad scope of the clauses he/she must take into consideration in preparing or reviewing a contract. Risks related to contract provisions are detailed to give contract managers a good understanding of this topic.

Chapter 6 reviews in detail time and costs related risks and liabilities. This chapter reviews, as well, delay and disruption claims and loss and expense components.

Chapter 7 summarises the main topics a contract manager must be familiar with. They are listed in alphabetical order and include dispute resolution areas such as adjudication, mediation and arbitration as well as the critical areas of suspension, termination and *force majeure.*

A note on the use of personal pronouns in this book: when used with regard to the parties to a contract (the contractor, the engineer …) these are interchangeable; so, for example, 'he' denotes 'he', 'she' or singular 'they'. Sometimes 'it' is used with, for example, 'the contractor', and should be taken as a reminder that the contractor may be a company rather than an individual.

Note also that the numbers given for days or times are indicative, too, and they can be changed according to the relevant contract as agreed between the parties.

Riyadh, Saudi Arabia Ali D. Haidar

References

Cook C (2014) Successful contract administration: for constructors and design professionals, 1st edn. Routledge, Abingdon

---

[1] AIA—The American Institute of Architects; DBIA—Design Build Institute of America, ConsensusDocs.

[2] ICE—Institution of Civil Engineers Conditions of Contracts; NEC—New Engineering Contracts; JCT—The Joint Contracts Tribunal.

[3] FIDIC—The International Federation of Consulting Engineers.

Corey Jr J (2015) Contract Management and Administration for Contract and Project Management Professionals: A Comprehensive Guide to Contracts, the Contracting Process, and to Managing and Administering Contracts. Self Published

Contents

Abbreviations

AC	Air Conditioning Unit
AIA	The American Institute of Architects
BOQ or BQ	Bill of Quantities
BOT	Build-Operate-Transfer
D&B	Design and Build
DAB	Dispute Adjudication Board
DBIA	Design Build Institute of America
FIDIC	Fédération Internationale Des Ingénieurs-Conseils
GMP	Guaranteed Maximum Price
HSE	Health and Safety Executive
ICE	Institution of Civil Engineers
LEED	Leadership in Energy & Environmental Design
O&M	Operation and Maintenance
QA	Quality Assurance
QC	Quality Control

Chapter 1
Introduction to Contract Management

Abstract The construction process is complex and involves high levels of planning, detailed specifications, payment process and documentation procedures. This book's purpose is to address all these issues and to fill the gap in the general literature available by providing a comprehensive textbook that addresses most of the topics a contract manager or a professional in construction needs to be familiar with. A good example is Chap. 7 (A to Z in Contract Management), which basically addresses in alphabetical order the major topics and clauses that any professional in the construction industry needs to know about contract management. This book is a complete reference work for professionals in construction, and for students embarking on higher degrees in contract and project management, who wish to familiarise themselves with construction and contract law, with the types of construction contracts and procedures (such as lump sum, cost plus and BOT contracts), and with the essential skills required in drafting and implementing a contract. The book also reviews generally the standard forms of contract in use worldwide, and the main recurrent conditions and risk provisions a contract manager must understand.

Keywords Construction law · Contract law · Contract management · Contract administration · Standard forms of contract

Contract Management: Process

Contract management refers to the process and procedures that parties to a contract implement in order to manage the negotiation, execution, performance, modification and termination of contracts between contractors, subcontractors, suppliers, consultants and employers.

In most cases, a contract manager is responsible for drafting the contract and all related documents as well as identifying and resolving specific legal issues; hence, it is absolutely essential that he/she works closely with the different parties involved in a construction contract to establish well in advance their mutual expectations regarding the role that they will be expected to play in negotiating, drafting, finalising and monitoring a particular contract (Raina 2009).

A. D. Haidar, *Handbook of Contract Management in Construction*,
https://doi.org/10.1007/978-3-030-72265-4_1

Project and construction managers will typically be responsible for identifying and resolving the operations and the risk management issues associated with the contract and the underlying relationship between the parties. A contract manager can become heavily involved, however, in the negotiation of commercial issues, and can have a great deal of input into the strategy goals and objectives of a contractual arrangement.

Here are a few illustrations to highlight the process of contract management:

1. Any construction project unfolds in phases: conception, design, pre-construction, procurement (securing materials, equipment and work teams), building and delivery/post-construction. Contracts cover pretty much all of them, and can be broken down into pre-award and post-award contract development.
2. In complex projects, contract changes will occur, and need to be managed usually by the construction management team. Ideally, this process is included in the original contract documentation.
3. The construction management team must have the knowledge to formulate and interpret the essential contract clauses that are needed for the different phases of a project. Contract managers assist in formulating the details of the contract arrangement, working with prospective partners to negotiate contract matters such as costs, specifications, schedules and performance. They also deal with other issues that are also imperative in a contract, and which include site handover, design, payment terms, suspension, termination, *force majeure* and others as described in this book.
4. This expertise is invaluable when it comes to minimising risks of all parties involved.
5. Contract management requires the appropriate utilisation of the designated party resources to meet its operations objectives.
6. A contract manager must, also, be familiar with bonds, guarantees and warranties and their relationship with indemnity, the responsibilities of the parties, and the basic doctrines in law related to contract and construction law.
7. Contract management also entitles liabilities after a contract has been completed and taken effect. Thus, this entails working to ensure that the terms and conditions contained within the contract are adhered to, and that each party's contractual obligations are met satisfactorily (Davidson 2015; Prasad 2010).

These facets of the contracting cycle are essentially administrative and contractual in nature and require a great deal of strategy and awareness of the different aspects and technicalities involved in a project.

In addition, during the contract management phase, circumstances will change, necessitating modifications to the contract.

The contract management team tends to work closely with the other parties to the agreement, so they are in a good position to know whether the parties have deviated from the contract in place, and whether they are complying with the express and implied terms of the contract.

While drafting a contract, the employer must ensure that the clauses used in the documents are clear, accurate and most importantly fair and reasonable.

If a contract is clear and unambiguous, its effect cannot be changed merely by the course of conduct adopted by the parties. Therefore, the clauses in the contract should be capable of being abundantly clear by the other party to a contract (Hughes et al. 2015).

Objectives

The objective of the development of any contract is simple: a clear understanding on the part of both parties as to their responsibilities, and a fair balance of risks and opportunities. The objective of contract management is equally simple, that is, to ensure compliance with the contract (Corey 2015).

If the project objectives are to be met in a competitive environment, the difficulties of achieving this are considerable and include:

- redrafting established clauses;
- addressing escalating prices;
- uncertain costs;
- milestone dates and increased risks;
- the failure of a party for non-delivery of the project;
- a failure to address the basic works, which the parties to the contract must re-address.

The process of drafting a contract forces both parties to evaluate what is expected of each other in the relationship, and establishes the foundations so that if things go wrong, there is a mechanism to resolve the issues and disputes that arise.

There are many ways of structuring a contract and getting contracts in place, with each option offering different balances of performance. The contract manager must ensure that the final contracting strategy decided upon aligns with the key performance requirements of the project. When this is in place, not only is the project more likely to achieve its goals but also the risks are properly understood and managed (Faiz 2017).

Contract Manager Role

One cannot overemphasise the importance of determining the contract manager's role in the contracting process, and the level of active involvement that he/she may have in negotiations relating to the contract,[1] negotiating and eventually drafting the contract.

[1]Even if the contract manager is not expected to play a primary role in negotiating the terms of the contract, he/she should make every effort to encourage managers to notify them as soon as possible that the transaction is contemplated.

It is not necessary for the contract manager to become a lawyer, nor even an expert in law. The contract manager needs, however, to have an awareness of the legal concepts in order to be able to effectively use the resources of the company's legal specialists to put the details into place (Carter et al. 2012).

In short, the contract manager needs to know the applied fundamentals of contract and procurement management that are applicable, and when to get the legal department involved. If a contract is set up and managed correctly, there will be minimal need to get the lawyers involved.

In the current state involving joint ventures, partnerships, multiple contractors, specialists and nominated subcontractors it may not be clear exactly with whom a party has a contract. Therefore, the contract manager must understand the works involved, the size of the project, its duration, the risks involved, and the capacity of the parties before embarking on a specific task. Contracting relationships are only successful where there is an equitable balance of risks and opportunity. This may be achieved through the right choice of pricing in the contract (for example lump sum, remeasured, cost plus etc.) and change orders mechanism (Prasad 2010).

Skilled contract managers understand that successful contracting involves all the considerations discussed herein. Above all, they understand that contracts are formed with and managed by people, and interpersonal skills are critical in developing a successful contracting relationship management process, regardless of their role.

In all situations where the contract manager is empowered to take some actions without other managers being present, he/she should make sure that procedures are in place to promptly communicate any new development to the appropriate staff within the designated party.

Contract management involves working with the appropriate representatives of the designated parties and involves (Faiz 2017):

- identify the steps that need to be taken in order to comply with the requirements of any contract review;
- signature authority policies;
- procedures that have been established by the designated party. For example, does the contract need to be reviewed and approved by senior management and/or the board of directors? If so, consider what needs to be done in order to expedite review and consideration;
- understand the scope of the proposed contractual relationship;
- identify the contract documents required to document the relationship;
- proceed with collecting and reviewing examples of the necessary contracts to expedite the drafting process;
- isolate specific questions that the designated party will need to answer for the contract to be complete and accurate;
- prepare a time and responsibility schedule for drafting, review, discussion, revision and completion of all clauses and required items and activities;
- consider discussions with designated party representatives regarding the contract manager role;
- participate in the negotiation of the essential terms of the contract; and

- prepare a term sheet or letter of understanding to be sure that the parties agree upon the essential terms before time and effort is spent on contract preparation.[2]

Once background information has been collected and preliminary agreement has been reached regarding the essential terms, the contract manager should prepare the initial draft of each of the required contract clauses and related documents. In cases where the opposite party is responsible for drafting, the contract manager's responsibility includes reviewing the initial draft of such items prepared by the opposite party and negotiating necessary changes in the initial drafts to make sure that the revised drafts are forwarded to the responsible parties for review and finalisation.[3]

Once the documentation is finalised, the contract manager should prepare closing of the transaction, including pre-closing meetings and preparation of the final check-lists and memoranda. If certificates and/or consents from outside parties are required for the contracts to be finalised and become effective they must be planned well in advance and may themselves require time-consuming negotiations.

Once the closing is completed the contract manager should make sure that all the documents are organised and that copies are delivered to the interested parties. This is also the time for the contract manager to make sure that the files relating to the transaction are organised so that they can be easily accessed in the future if needed.

Contractual Arrangement

When one is setting up a contractual agreement for a construction project, a large number of factors need to be considered for inclusion in the contract. In summary, these factors could be grouped into three main categories:

- Applicable law. A construction project must be built in compliance with laws and regulations. Such laws and regulations could be a regional planning program, national building codes, national laws and international regulations (Uff 2009).
- General and special terms and conditions. The terms and conditions of a contract contain a predefined standard form of contract, which must be included in any construction contract. They are applicable to every construction project.

[2]If the contract manager is not to be directly involved in negotiations he/she should, at a minimum, provide the designated party representatives with a list of questions that will need to be answered in order for the contracts to be completed so that the representatives can discuss them with their counterparts from the opposite party.

[3]The timing of the drafting and revision process is crucial since delays can push the relationship off track and jeopardise realisation of the business opportunities anticipated by both parties.

- Project particulars. Each project has different characteristics regarding location, soil conditions, environment, type of construction (e.g. residential building, industrial plant, roads, airports) and the employer (e.g. private/public, financial power, quality standard).

Every contract has the same fundamental elements, however. There must be an offer and acceptance. In a two-party relationship, the first party provides an offer (performance of work, service) and the second party has to accept it. Upon signing, the intentions of the parties are bound by the contractual terms, and the contract comes into force (Haidar and Barnes 2017).

In the construction industry, there are many types of contracts that are in effect. Depending on the type of employer (public, private) and the type of the project organisation, the following types of contracts exist generally in a construction project (Davidson 2015):

- Financing contract (employer ↔ outside investor).
- Consulting contract (employer ↔ external engineering specialist).
- Design contract (employer ↔ designer).
- Engineering or construction works contract (employer ↔ contractor).
- Subcontract (contractor ↔ subcontractor).
- Delivery contract (contractor/subcontractor ↔ supplier).

Each standard form of contract has a similar content; for example, a civil works contract is divided into two different parts, a legal part (general, additional and special[4] conditions) and a technical part (general specifications, technical specification, additional technical specifications).

The following terms are essential in a works contract and must include an exact specification of each of them:

- Information about the project and contractual parties.
- Duties and responsibilities of the employer.
- Duties and responsibilities of the contractor.
- Specification of work—scope of services.
- Reimbursement.
- Liability.
- Distribution of risks.
- Terms and deadlines.
- Change of order/ interruption of work.
- Acceptance of work.
- Payment terms.
- Suspension and termination .
- Guarantees/warranties.
- *Force majeure* (Raina 2009).

[4]Also called 'particular conditions'.

For the construction phase to commence, with the risks minimised so that the employer and the engineer can manage the works up to the delivery and commissioning phase, the following points should be observed (Prasad 2010):

- Take steps to resolve differences in the interpretation of the output specification.
- Monitor the progress of project delivery and the quality of work.
- Supervise the conduct of required tests, evaluate the test results, and take decisions as required.
- Consider variations to the output specifications.
- Inspect equipment and materials to be used.
- Certify and provide approvals as may be needed under the contract.

Construction and Contract Law

Construction projects are unique in nature, and every situation and every problem in construction is different and contains unique facts that may require a different approach and solution to one that would be appropriate in another circumstance. Increasingly, common law is modified by statute, and the services of legal experts are often required to provide legal advice and guidance on specific issues and to assist the parties to a construction contract their legal status in case of a dispute.[5]

A construction contract can be described as: 'an entire contract for the sale of goods and work and labour for a lump sum price payable by instalments as the goods are delivered and the work done'.[6]

In many ways, construction law is no different to any other field of law; however, it is recognised as being a body of law that relates to those elements of the law that directly affect the construction and civil engineering industries.

Construction law covers legal and semi-legal topics and doctrines such as contract law, the law of tort, construction claims and most importantly disputes resolution. It is particularly concerned with the effect that these elements of the law have upon the employers, consultants, contractors and subcontractors involved in the construction process (Uff 2009).

The basic principles of construction law involve:

- clearly defining and identifying the responsibilities of the contracting parties;
- defining notice periods and time bars required to protect the parties against prejudice or error;
- setting standard methodology for dealing with contractual responsibilities and obligations;
- promoting contractual protection of the innocent party in the event of default;
- establishing practical options within the contracting process;

---

[5]Decisions have to be made from time to time about such essential matters as the making of variation orders, the expenditure of provisional and prime cost sums, and extension of time for the carrying out of the work under the contract.

[6]*Modern Engineering (Bristol) Ltd v Gilbert-Ash Northern (1974) AC 689.*

- providing for reciprocal guarantees between the contracting parties; and
- setting payment conditions that offer significant protection to the contractor and subcontractors; (Hughes et al. 2015).

A contract can be considered as being an agreement that gives rise to obligations that are enforced or recognised by law. Therefore, the essence of any contract is agreement. In deciding whether there has been an agreement, and what its terms are, courts usually look for an offer to do or to forbear from doing something by one party and an unconditional acceptance of that offer by the other party—turning the offer into a promise.

In addition, the law requires that a party suing on a promise must show that he has given consideration for the promise, unless the promise was given by deed. Further, it must be the intention of both parties to be legally bound by the agreement, the parties must have the capacity to make a contract, and any formalities required by law must be complied with (Uff 2009).

If there is fraud or misrepresentation the contract may be voidable, while if there is a mutual mistake about some serious fundamental matter of fact this may have the effect of making the contract void.

Finally, there must be sufficient certainty of terms and, partly due to this, in the construction industry, standard forms of contract are regularly used.

The parties to a construction contract are bound to each other for a certain period of time by a unique and exclusive relationship they created for their mutual benefit. This unique relationship, called 'privity of contract' gives them both obligations and rights which they have agreed to accept so that both may benefit (Haidar 2011).

This contractual relationship persists until the contract is discharged or terminated, that is, until it is performed, or terminated because of impossibility, agreement (by the parties), bankruptcy (in some cases), or breach of contract.

The basic principles of construction and contract law are detailed in Chap. 2, where the most relevant doctrines are described.

Standard Forms of Contract

Standard forms of construction contract seek to regulate the relationships between the contracting parties, particularly in respect of risk, management and responsibility for design and execution.

It would, however, be practically impossible to devise a standard form of contract that would provide for all eventualities in a construction project, as there are several factors that affect what type of contract is suitable for a certain project, e.g. the amount of involvement from the employer, the technical complexity, and the location, nature and size of the project (Baker et al. 2013).

The main advantages of using standard forms of contract are that they are usually generated by the different bodies that make up the interests of standardisation and good practice in the construction industry. As a result, the contractual risks are spread equitably.

In addition, using a standard form precludes the cost and time of individually negotiated contracts, and tender comparisons are made easier since the risk allocation is the same for each tenderer.[7]

The disadvantages are that the forms are cumbersome, complex and often difficult to understand; thus the resulting contract is often a compromise between the parties to reach a common document they can agree to.

Standard forms of contract are usually referred to as the general conditions. There is no rule as to what should be included in the general conditions of a contract, but, according to most sets of conditions, they follow a standard pattern. Typically, the conditions deal with:

- General obligations to perform the works.
- Provisions for instructions, including variations.
- Valuation and payment.
- Liabilities and insurances.
- Provisions for quality and inspections.
- Completion, delay and extension of time.
- Role and powers of the employer representative or the engineer.
- Disputes.

Amendments to standard contracts should be kept to a minimum. Increase in the number of supplements to standard forms of contract will decrease the effectiveness of a standard form and can result in misinterpretation and disputes later during the execution of the contract works.

Some employer/contractor organisations develop their own 'in house' set of terms and conditions for use on their projects. Such terms and conditions of contract are referred to as bespoke conditions of contract.

Contract Documents

An exhaustive construction contract can include a number of different documents, all of which define responsibilities and risks regarding the different aspects of the project. Some of these documents are the general conditions, the special conditions, specifications, drawings, bill of quantities, templates for subcontractor agreements, warranties and assignment and general forms for bonds and guarantees (Haidar 2011).

The question necessarily arises as to whether these documents fit together, which (if any) are to have precedence, and what will happen if they are in conflict. There are two distinctly different approaches to these questions. The first, and simplest

[7] Parties are assumed to understand that risk allocation and their pricing can be accurately compared.

approach, is to make all contract documents of equal weight and significance. Another solution sometimes found is to provide that the contract documents shall have an order of precedence, i.e. a conflicting requirement in two documents is to be resolved in favour of the one having the higher priority.

The contract documents shall be deemed to be mutually explanatory of one another. In the event of ambiguity, discrepancy, divergence or inconsistency in or between them, this agreement shall prevail over all other contract documents.

References

Books

Baker E, Mellors B, Chalmers S, Lavers A (2013) FIDIC contracts: law and practice. Taylor and Francis, Milton Park

Carter R, Carter A, Kirby S (2012) Practical contract management. Cambridge Academic, Cambridge

Cook C (2014) Successful contract administration: for constructors and design professionals, 1st edn. Routledge, Abingdon

Corey Jr J (2015) Contract management and administration for contract and project management professionals: a comprehensive guide to contracts, the contracting process, and to managing and administering contracts. Self published

Davidson AC (2015) Contract management: a contractor's perspective. Createspace Independent Publishing Platform, Scotts Valley CA

Faiz S (2017) Basics of construction contracts. Routledge, Abingdon

Haidar AD (2011) Global claims in construction. Springer Verlag, London

Haidar AD, Barnes P (2017) Delay and disruption claims in construction. ICE Publishing Ltd, London

Hughes W, Champion R, Murdoch J (2015) Construction contracts law and management, 5th edn. Routledge, Abingdon

Prasad L (2010) Managing engineering and construction contracts: some perspectives: revisiting the segments of contracts management. LAP LAMBERT Academic Publishing, Saarbrücken

Raina VK (2009) Raina's construction and contract management, 2nd edn. Shroff Publishers and Distributors Pvt. Ltd, Mumbai

Uff J (2009) Construction law, 10th edn. Sweet and Maxwell, London

Case law

Modern Engineering (Bristol) Ltd v. Gilbert-Ash Northern (1974) AC 689

Chapter 2
Construction Contract Law

Abstract This chapter is intended for contract managers and professionals involved in project and contract management to provide them with a basic understanding of the nature of the law as it applies to contracts and doctrines. Doctrines in contract formation are needed in order to deal with the considerable complexity and interrelationship between the contractor, the employer and the consultant as well as the specialists and the subcontractors. The chapter will look at the principles in contract formation such as agreement, offer and acceptance, and consideration. Doctrines and concepts in construction contract law such as terms of an agreement, letters of intent, parties to a contract and general doctrines such as misrepresentation, mistake and frustration are included. The objective is to make the reader understand these rules in order to judge the validity of the contractual terms and to assess if a contract is concluded between the parties.

Keywords Contract formation · Contract law · Construction law · Breach of contracts

Understanding Law: General

There are real differences in focus, terminology and legal definitions between the various branches of law. Lawyers and judges have adopted several main classifications in categorising law, and these classifications play an important role in how the law is written and practised. They are as follows:

1. Laws that deal with contract, tort and property. They also deal with rules or principles and remedies derived both from common law and from equity.
2. Criminal law, which is also called penal law. Criminal law involves prosecuting a person for an act that has been classified as a crime. Civil cases, on the other hand, involve individuals and organisations seeking to resolve legal disputes.
3. Substantive law. This includes the rules that lay down rights and duties and enable individuals to modify their legal position.
4. Law of procedure, which defines how rights and duties can be given effect in civil or criminal proceedings.

A. D. Haidar, *Handbook of Contract Management in Construction*,
https://doi.org/10.1007/978-3-030-72265-4_2

5. Public law (as opposed to private law), which regulates the relations between individuals and other private legal persons in relation to the powers of local authorities and other public bodies.

The sources of law are forces that make the law how it is. They encompass socio-economic influences, the political process, the background and education of the judges, the behaviour of litigants and their advisers, and how cases get to court (Campbell 2019).

The sources of law are the categories of material that the courts regard as authoritative in deciding the cases before them. This is a definition that focuses on the courts, which in normal circumstances have the last word on what the law is, subject only to overturning by a higher court or by some higher source of law, such as Parliament or a similar authoritative body.

Doctrines in law always come from a recognised and legitimate source. These sources are utilised in order of importance:

1. *Statute* is an express and formal laying down of a rule or rules of conduct to be observed in the future by the parties to whom the statute is expressly, or by implication, made applicable. A statute gives no reasons and is imperative.
2. *Case law*, usually called common law, takes the form of court judgments. A judgment gives reasons and may be argumentative.
3. *Regulations* include building and environmental regulations, health and safety procedures and other government bodies' formulation of rules and regulations.
4. *Civil bodies* such as institutions for engineers, architects, lawyers, civil servants and other professionals. Standard forms of contracts fall under this category.
5. *Publications* in the form of books, papers, journals, and interviews with qualified professionals or persons of law.

An Overview of Construction Contract Law

Contract law, as opposed to tort, is civil rather than criminal unless injury or accidental death should occur. The theoretical division between contract and tort is that liability in tort arises from the breach of an obligation primarily fixed by law, whereas in the contract it is fixed by the parties themselves.

The law of contract is frequently the first 'case law' subject to which students are introduced when they commence their legal studies. The main reason for this is that contracts affect the general public more than most other areas of law, and arise daily in business and commercial life. The contract is the most important stage in the process when land or buildings are transferred and when building projects are undertaken (McKendrick 2009).

A *prima facie* duty of care arises if there is sufficient proximity between the alleged wrongdoer and the wronged party such that the former might reasonably expect that carelessness may cause damage to the latter. It is then necessary to consider whether there are any mitigating circumstances that reduce or limit the scope of the duty and damages.

The period of time after which a cause of action starts to run in a contract is from the moment when the conduct constitutes a breach, whereas in tort it starts from the moment when the plaintiff sustains his damage, which is usually at a much later stage.

There is no single correct way to divide up general law for the purpose of breaking it down into comprehensible components. Looking at it by its source makes little practical sense, except perhaps in relation to the distinction between law and equity. It makes more sense to use a coherent factual subject area, like construction (contract) law (Richards 2009).

There is, however, no authoritative definition of what construction law covers or includes, so legal categories are a matter of choice; hence it is a matter of finding the appropriate choice that is informative and appropriate for the purpose to be achieved.[1]

Doctrines in construction contract law and subsequently the implementation of these laws in contract formation are needed to set the boundaries of proper acceptable behaviour between the parties to a contract, to save time in agreeing to the principles of an agreement and to assist in dealing with common contractual situations and reoccurring problems and disruptions. They also handle the unexpected risks in the industry and deliver the required level of certainty to the whole process.

They are also necessary in order to deal with the considerable complexity and interrelationship between the contractor, the client and the consultant as well as the specialists and the subcontractors. This complexity arises from:

1. The split responsibility for design and then construction.
2. The large number of independent subcontractors.
3. The complex price determination and procurement.
4. The large variety of specialists and suppliers involved.
5. The infinite number of transactions necessary to deal with the very large projects that are most common nowadays.

A contract is made up of a set of mutual promises stating the rights and obligations of the parties to that contract. These rights and obligations become enforceable by law and, therefore, parties rely on contracts in structuring their business relations.

Although parties have significant flexibility in setting the terms of their contracts, the enforceability of these terms is subject to limits imposed by relevant legislation and common law.

Contracts are, therefore, private law created by the agreement between contracting parties within the parameters of well-established basic legal mechanisms.

[1] Construction contract law, together with its many contents and assuming that it covers the ground adequately, deals with all the primary sources of law.

Formation of a Contract

Whatever contract may be used, the actual works on almost all construction projects exhibit recurrent distinctive characteristics:

- The prototypical nature of the works.
- Split responsibility for specification and design.
- High degree of interactivity between client, contractor and supplier.
- Expectation of, and provision for, substantial levels of change.
- Scope of work, complexity of sequencing of activities, and dependencies on other activities or supplies.
- Site specificity.
- Exposure to, and dependence on, outside factors such as international fluctuations in raw materials prices, lack of resources due to demand, and Acts of God.
- Longevity of the products, and lateness of revelation of defects.
- The diversity and sheer volume of evidentiary material.

The essential elements of the formation of a valid and enforceable contract can be summarised under the following headings (Beatson et al. 2010):

1. There must be an offer and an acceptance which is in effect the agreement.
2. There must be an intention to create legal relations.
3. There must be consideration.
4. There must be validity of contract.

The significance of a contract is that the promisee is entitled to performance of the promise or, failing that, to compensation for non-performance. A contract by law[2] is enforced in two stages:

1. The first stage is decision, mostly on whether there is a right to payment or compensation, and how much.[3]
2. The second stage is that a court can enforce a judgement or arbitration award through various processes such as the appointment of a receiver, seizure of goods, etc.

Where the law provides a framework within which the activities of the construction industry are carried out, a contract provides a sub-framework of the law for a specific undertaking. The contract arises from the agreement between two or more parties and it binds those parties in a contractual arrangement.

The requirements and related documents of a project are important in determining whether a contract exists. In some situations, one party or another may seek to prove that either there is or there is not a contract, in order to secure a benefit or avoid a liability.

---

[2]The court will rarely enforce performance, other than payment, because it cannot or will not supervise actual performance of work.

[3]The law empowers the courts and arbitrators to make a binding decision on disputes properly submitted to them.

Offer and Acceptance

Although it is not essential to do so, it is conventional to analyse an agreement in terms of offer and acceptance.

The person making the offer is called the offeror and the person to whom it is made is called the offeree.

These principles are used not only to decide when and whether a contract has come into existence, but also what terms have survived and emerged as part of the contract. The general rules relating to offer and acceptance are:

1. A standing offer can be revoked at any time only by making the offer and only before acceptance by another.
2. The effect of an offer is that it confers on the offeree the power to accept it and, thereby, to create a binding contract.
3. An acceptance outside the time for valid acceptance will be too late. If no time is specified, the acceptance must be made within a reasonable time.
4. The ordinary rule at common law is that an offer is accepted when the offeror receives notice of the acceptance.

It is also determined that introducing a new term, or deleting a term, is a counter-offer. The effect of the counter-offer in law is to kill the original offer so that it cannot then be accepted unless it is revived. But a mere request for further information is not a counter-offer.

This principle enables the outcome of a sequence of exchanges to be analysed, both to determine when and whether a contract has come into existence and what the outcome is.

The main principles of an offer are:

- It is a proposal by one party to enter into a legally binding contract with another.
- It may be oral, in writing or implied by conduct.
- It must be promissory in nature.
- It must be sufficiently complete (that is, it must contain essential terms).
- It must be intended to result in a contract if accepted.

It must be addressed to a particular party or identified group of persons, or to the world at large.[4] The main principles of acceptance are:

- An indication of readiness to contract on the offered terms.
- Accepted by the offeree while offer in existence.
- Accepted in reliance of the offer.
- Accepted in the same terms as the offer;
- Accepted unconditionally.
- Acceptance must be communicated to be effective.

An offer may be terminated in the following ways:

---

[4] *Carlill v Carbolic Smoke Ball Company [1892] EWCA Civ 1.*

1. Revocation:

- This must be communicated to the offeree before the offer is accepted.
- An offer can only be revoked on expiry.

2. Rejection:

- If the offeree rejects an offer, he cannot accept it later.
- Acceptance on different terms is actually a rejection and the making of a counter-offer.

3. Lapse of time:

- Non-acceptance within a specified or reasonable time.

Intention to Create Legal Relations and Essential Terms

In determining whether parties have created legal relations, courts will look at the intentions of the parties. If, in the course of the business transactions, the parties clearly and expressly make an agreement stating that a contract ought not to be binding in law, then the courts will uphold those wishes. However, if a court is of the view that there is any ambiguity of intention, or that such an intention is unilateral, then the contract will be void.

In construction agreements, there is a rebuttable presumption that the parties intend to create legal relations and conclude a contract.

Another requirement for a legally binding contract is agreement of the essential terms. This has two main elements—what an agreement is, and what its essential terms of an agreement are.

The old legal concept of *consensus ad idem*, a meeting of minds, has been replaced by the objective approach that an agreement is not a mental state but an act, and, as an act, is a matter of inference from conduct where the parties are to be judged, not by what is in their minds, but by what they have said or written or done.

The law requires that for an agreement to be a binding contract, not only should all essential terms be agreed, but also the terms agreed should be sufficiently certain. The courts will not make an agreement for the parties: rather they will try to give meaning to a contract if they find the parties intended to be bound by it. Repugnant or surplus words may be rejected or modified, or clearly intended words supplied, in order to save a document.

Consideration

The third requirement for a binding contract is that a promise is only binding if it is given for good consideration, or if it is executed as a deed. Under law, an agreement supported by consideration is not enough to create a legally binding contract; the parties must also have an intention to create legal relations.

Valuable consideration has been defined as some right, interest, profit, or benefit accruing to one party, or some forbearance, detriment, loss, or responsibility given, suffered, or undertaken by the other at his/her request.

The classic definition of consideration was provided in a statement by Sir Frederick Pollock, adopted by Lord Dunedin in Dunlop Pneumatic Tyre Co Ltd v Selfridge & Co Ltd [1915] AC 847: 'An act or forbearance of one, or the promise thereof is the price for which the promise of the other is bought, and the promise thus given is enforceable'.

The presence of consideration is often indicative of the intention to create legal relations, though there are situations where the presumption of the intention can be rebutted, thus determining that there is no contract and no legal liability.

It is not necessary that the promisor should benefit by the consideration. It is sufficient if the promisee does some act from which a third party might benefit, and which he would not have done but for the promise.[5] Therefore, consideration for a promise may consist in either some benefit conferred on the promisor, or detriment suffered by the promisee, or both. On the other hand, this benefit or detriment can only amount to consideration sufficient to support a binding promise where it is causally linked to that promise.

The purpose of the requirement of consideration is to put some legal limits on the enforceability of agreements even when they are intended to be legally binding and are not vitiated by some factor such as mistake, misrepresentation, duress or illegality.

The traditional definition of consideration concentrates on the requirement that something of value must be given, and accordingly states that consideration is either to the detriment of the promisee in that he may give something of value, or of some benefit to the promisor in that he may receive a certain value. Law is concerned with consideration for a promise, not consideration for a contract.

To enforce a promise, a promisor must show that he/she has given something or promised to give something in return for the promise (i.e. given consideration for the promise):

- It is the price paid by one party in exchange for the promise of the other.
- Consideration arises from the notion that a contract is a bargain struck between the parties by an exchange.
- Consideration must be present for a 'simple' (informal) contract to be enforceable (formal).

---

[5]Currie v Misa (1875) *LR 10 Exch 153.*

Validity of Contract

In order for a valid contract to come into existence, certain requirements must be satisfied. These requirements are (Kelleher and Walters 2020):

- The parties should be competent to contract.
- There should be agreement (consensus) between the parties.
- The contents or consequences of the contract should be possible at the time that the contract is entered into.
- The subject matter of the contract should be legal.
- The contract should not be contrary to public policy.
- The contract should be voluntarily, seriously and deliberately entered into for some reasonable cause.
- The formalities required by law should be observed.

Implied Terms

In analysing a contract, a lawyer identifies the terms of that contract. The meaning of this can be explained thus: 'if a statement is a term of the contract, it creates a legal obligation for whose breach an appropriate action lies at common law'.

In the event of a breach of a term of a contract the basic right which derives from the contract is to bring an action, that is, to sue upon the breach and recover money damages in respect of loss or damage suffered as a result of the breach.

Terms may be classified according to a number of characteristics. The first is that terms may be expressed or implied.

Express terms are those terms to be derived from the written or spoken words, actions and/or conduct forming the agreement.

Terms are implied by law to supplement the express terms. Standard terms will be implied to avoid the need for the parties to express each time that which goes without saying. Specific terms may be implied to fill in gaps in the parties' agreement.

The implication of terms in specific transactions is governed by different principles. A term will be implied, where necessary, to give business efficacy to the agreement and to give effect to the presumed intentions of the parties. A term will readily be implied in a contract so that the parties shall cooperate to ensure the performance of their bargain. However, the legal concept of cooperation is restricted to an obligation firstly to perform those duties which must be performed to enable the other to perform his own obligations under the contract and secondly not to prevent the other from performing.

The basis on which the courts can imply terms is controversial. Traditionally, the justification has been that they were giving effect to the presumed intention of the parties. In such cases, the idea is that it is necessary to presume that the term was intended; otherwise, the contract would not function properly.

It is possible to divide terms incorporated by the courts into two categories: terms implied in fact and terms implied in law. In the latter case, the term will be implied in every contract of that kind, whereas in the former case, the term will only be implied where the facts of the case give rise to a need, i.e. where there is evidence that it was an unexpressed intention of the parties. The importance of the distinction is that the test seems to be lower for a term to be implied in law, where it is only reasonably necessary, whereas for terms implied in fact it must be necessary.

Privity

The general concept of a contract involves obligations voluntarily undertaken towards specific parties. This concept was developed restrictively into the doctrine of privity of contract, whereby a party that is not part of a contract can neither sue on, nor rely on a defence based on, that contract unless it effects public liability.

In other terms, the doctrine of privity means that a contract cannot, as a general rule, confer rights or impose obligations arising under it on any party other than those that are the parties to it.

The law has a number of doctrines relating to the capacity of certain classes of a party to enter into binding contracts. There are various doctrines concerning the question of whether contracts made by such parties are void, voidable or unenforceable by one side or the other.

As to the principles of privity, in common law there is the principle that consideration must move from the promisee[6] and that only a promisee may enforce the promise, meaning that if the third party is not a promisee he is not privy to the contract.[7]

The implications of privity are significant in construction. For example, subcontractors are not parties to the main contract between employer and main contractor, so that the employer cannot sue the subcontractor for breach of contract, and vice versa. The theory is that rights and obligations should be pursued up and down the contractual chain. There must be, however, an intention to create a collateral contract before that contract can be formed (Furmston et al. 2006).

Novation and Assignment

An assignment (*cessio*) encompasses the transfer of rights held by one party, the assignor, to another party, the assignee.

The topic of assignable rights has been the subject of intense judicial and academic consideration. It pitches against each other two fundamental principles: namely,

[6]Tweddle v Atkinson (1861) *1 B&S 393*.

[7]*Dunlop Tyre Co v Selfridge* 1915.

freedom of contract and freedom to dispose of one's property. The collision of these two principles is compounded by long-standing difficulties of characterising assignment in terms of the principle that assignment is an exception to privity of contract.

The burden of a contract cannot be transferred without the consent of the other party. If a new contract is substituted with the consent of the other party, it is called a novation; the new contract then refers back to the commencement of the original contract.

The benefit of a contract may be assigned to a third party without the consent of the other contracting party. If this is not desired, it is open to the parties to agree that the benefit of the contract shall not be assignable by one or other of them, either at all, or without the consent of the other.

Certain expressed terms in special construed contracts conditions, providing for a prohibition against assignment of obligations under the contract, are not usually contrary to public policy, and a purported assignment in breach of such a condition is ineffective.

Conditions that are precedent to an assignment are:

1. The construction contract is not subject to any claim, setoff or encumbrance.
2. There have been no prior assignments of the construction contract.
3. The construction contract constitutes the valid and binding obligations of the parties thereto, and is enforceable in accordance with its terms.
4. Neither borrower nor contractor is in default under the terms of the construction contract.
5. All covenants, conditions and agreements have been performed as required by the construction contract, except those which are not due to be performed until after the date of this agreement.
6. The assignee continues to be liable for all obligations of assignor under the construction contract.
7. The assignee, thereby, agrees to punctually perform and observe all of the terms, conditions and requirements of the construction contract to be performed or observed by the assignor.
8. The assignee agrees to indemnify and hold the assignor harmless against and from any loss, cost, liability or expense (including, without limitation, reasonable attorneys' fees, court costs and investigation expenses) resulting from any failure of the assignee to perform its obligations under the construction contract.

Letters of Intent

In an ideal world, all contracts would be properly drawn up and signed before work commenced, whether the contracts were between the employer and the main contractor or between the main contractor and the subcontractor.

It seems that in ever-increasing instances this is not the case. Fast-track construction methods often overtake the procedures for drawing up the contract, which in many instances lacks essential urgency. This applies in the construction sector and can often lead to disputes that prove time-consuming and expensive to resolve.

In many cases, however, the parties are anxious to make an early start even before all the key contractual matters have been agreed, and long before a formal contract has been drawn up. In the absence of a formal contract it is now common practice for the employer to send the contractor a letter of intent that allows the contractor to make a start on site or to commence design or order materials.

It is of limited advantage to a contractor or subcontractor to learn that he is entitled to a payment if there is no agreement as to how much the payment will be. From this decision it can readily be seen that even if a letter of intent includes a specific instruction to undertake work it does not necessarily mean that a contract has come into being (Allen and Martin 2010).

Therefore, it is common practice for the parties to sign the letter of intent but this in itself does not mean that a contract has come into being.

To establish that a contract has been concluded not only requires evidence of agreement by the parties on all the terms they consider essential, but also sufficient certainty in their dealings to satisfy the requirement of completeness. Letters of intent, as traditionally drafted, fail on both counts since they are usually incomplete statements preparatory to a formal contract that would come into operation.

When the parties fall out over the wording of a letter of intent, usually in relation to payment, the court has to decide whether the wording is sufficiently explicit to form a contract. This, in essence, provides that if the contractor carries out work as specified in the letter of intent the employer will make a payment. This is in no way a contract in the normal sense as it usually lacks the full details as to the extent of the work that is required to be undertaken, and all the terms that are to apply.

Unfortunately, while the letter of intent may include an instruction to undertake work, often it does not specify the manner in which payment is to be made. The parties are fairly relaxed as they are both under the impression that a formal contract will be produced reasonably quickly containing all the necessary terms and, therefore, that there is no reason for concern. It is only when a sticking point is reached, and no contract is agreed, that the parties begin to argue as to the manner and amount of payment to be made. The courts often resolve this problem by holding that payment should be made on a *quantum meruit* basis, which in essence means a sum that is merited by the work undertaken, or, in other words, a fair and reasonable sum. This only leads to another problem—the question of how much is a reasonable sum.

In the case of *British Steel Corporation v Cleveland Bridge (1984)*, the court held, *obiter dicta*, that if work is done pursuant to a request contained in a letter of intent it will not matter whether a contract did or did not come into existence, because if the party who has acted on the request is simply claiming payment his claim will usually be based on a *quantum meruit*.

Lord Justice Goff said:

> Now the question whether in a case such as the present any contract has come into existence must depend on the true construction of the relevant communications which have passed between the parties and the effect (if any) of their action pursuant to those communications. There can be no hard and fast answer to the question whether a letter of intent will give rise

> to a binding agreement; everything must depend on the circumstances of the particular case. In most cases where work is done pursuant to a request contained in a letter of intent, it will not matter whether a contract did or did not come into existence; because if the party who has acted on the request is simply claiming payment, his claim will usually be based upon a *quantum meruit*, and it will make no difference whether the claim is contractual or quasi-contractual. As a matter of analysis the contract (if any) which may come into existence following a letter of intent may take one of two forms – either there may be an ordinary executory contract, under which each assumes reciprocal obligations to the other; or there may be what is sometimes called an 'if' contract, i.e. a contract under which A requests B to carry out a certain performance and promises B that, if he does so, he will receive a certain performance in return [and pay] usual remuneration for his performance. The latter transaction is really no more than a standing offer which, if acted upon before it lapses or is lawfully withdrawn, will result in a binding contract.

Contractual Risks: A Legal Approach

The trend in construction contracts has been shifting in recent years to include more risk sharing provisions and to reallocate increasing numbers and proportions of risks to the employer. Some forms of contracts probably place even more risk on the employer, but these contracts have made big advances by dealing with time and money and sharing responsibilities and risks in an integrated way. This should help to reduce abuse, but it does place higher demands on the engineer and the architect, in particular, to address compensation events due to changed conditions promptly, and to act fairly and with due diligence when certifying payments (Haidar 2011).

There is a recent tendency for the courts to apply what might be called the *portia* approach, favouring strongly interpretations of contractual provisions that presume that the parties did not intend a gamble and therefore imposing a presumption of business common sense; in other words, courts are discouraging the idea of '*first try to get out of the clause, but next try to construe things to your advantage*'.

Sharing, structuring and minimising the impacts of risks can be summarised by the following:

1. First, there is a distinction between risks that can be considered as insurable and those that cannot. With insurable risks the object of the contract is to ensure that the risks are allocated entirely to the party required to be insured. Since insurance is generally on an indemnity basis, the insured must be liable before the insurers will pay.
2. Second, the various criteria suggested for deciding risk allocation policy are neither necessarily consistent nor readily compatible. For example, the extent of unforeseeable ground conditions is best controlled by the employer through pre-contract ground investigations, but the consequences of unforeseen ground conditions, when encountered, are best managed by the contractor.
3. Third, structuring contracts, for risk sharing rather than straightforward risk allocation, tends to open up the scope for abuse.

The principal contractual provisions dealing with allocation of non-insurable risks are categorised under a number of headings:

- Quantities. These relate to the difference between actual quantities executed and quantities in the bill of quantity.
- Engineer's instructions and standards. These relate to the bidding and the execution processes.
- Changes in laws, regulations and taxes.
- Unforeseen physical conditions including soil conditions, weather and existing and neighbouring infrastructure facilities.
- Impossibility and *force majeure*.

The huge matrix of requirements, constraints and responsibilities necessary for the delivery of complex large projects will need to be project-managed and programmed by effective management of the numerous interrelated variables.

The contractor's unique expertise, coupled with his local knowledge and his historical endeavours in delivering projects, even in the most challenging environments, will place him, usually, at the forefront in carrying the usual risks associated with the works. In summary, the contractor is responsible for achieving the following objectives:

1. to integrate the construction programme for all the project(s) due to be instructed including the main works, the logistics, permits and resources;
2. to apply the best modern management techniques to achieve the predetermined objectives of achieving aims at cost, time, and purpose-built quality;
3. to provide alternative solutions for the construction and fabrication, considering alternative plans in terms of LEED,[8] sustainability and green technology; and
4. to provide alternative solutions in order to ensure that the development scheme will be a successful scheme in terms of construction methods, re-usage of temporary facilities, and infrastructure works such as transportation, electricity and water resources.

The contractor, in order to make the project successful, optimise costs and make sure that risks are minimised, must also implement the following methodologies and steps:

1. project scope management to ensure that all the works required are to be completed according to the program in place and according to the costs envisaged;
2. project time management to provide an effective project schedule;
3. project cost management to identify needed resources and maintain budget control;
4. project quality management to ensure functional requirements are met; and
5. project human resource management for the development and effective employment of key personnel for the project, including government committees and managers.

[8]Leadership in Energy & Environmental Design.

The employer and the contractor have a duty to express contract provisions that are fair and reasonable without avoiding the duty of care responsibilities imposed on both of them.

There has been an unspoken assumption that the contractor should be left to manage the performance of the works based on the design available and that certain categories of events are regarded as giving the contractor entitlement to extension of time for completion, but no financial compensation. This factor is, however, affected by certain mitigated factors in order to subdue the effects of these delays. Trying to shift the risk from one party to the other or in case one party tries to impose its rules, regulations and procedures on another can be disfavoured by the courts, especially if certain doctrines apply, as will be discussed in this chapter (Haidar 2011).

Project managers have an interest in risk management that should extend to knowing the strengths and weaknesses of the project management approaches if a dispute situation should arise. Changes to the schedule of works caused by clients and third parties often become contentious issues, especially when a change results in delay and disruption.

Professional Obligations in Construction

It is an accepted point, in fact and in law, that the contractor comes into a relationship with the engineer that is the result of the contractor entering into the contract with the employer and of the engineer having been engaged by agreement with the employer to perform the functions required under the contract.

The engineer assumes the obligation, under its agreement with the employer, to act fairly and impartially in performing its functions. The engineer is under a contractual duty to the employer to act with proper care and skill.

The contract provides for the correction, by the process of arbitration or by the courts, of any error on the part of the engineer, if indeed there is any real scope for an error on the part of the engineer that would not be at once detected by the contractor. The court, at least in the absence of any factual basis for the engineer to have foreseen any other outcome, will proceed on the basis that the contractor will recover the sums that it ought to recover under the contract (Campbell 2019).

It is foreseeable that a contractor under such an arrangement may suffer loss by being deprived of prompt payment as a result of negligent under-certification or negligent failure to certify by the engineer and in the case, among others, that the engineer does not provide an extension of time due to unforeseen conditions for which the contractor is not liable.

The contractual duty of the engineer to act fairly and impartially, owed to the employer, is a duty in the performance of which the employer has a real interest. If the engineer should act unfairly to the detriment of the contractor, claims will be made by the contractor to get the wrong decisions put right. If court proceedings are necessary, then the employer will be exposed to the risk of costs in addition to being ordered to pay the sums which the engineer should have allowed (Beale 2010).

If the decisions and the advice of the engineer, which caused the proceedings to be taken, were shown by the employer to have been made and given by the engineer in breach of the engineer's contractual duty to the employer, the employer would recover its losses from the engineer. There is, therefore, not only an interest on the part of the employer, in the due performance by the engineer to act fairly and impartially, but also a sanction which would operate, in addition to the engineer's sense of professional obligation, to deter the engineer from the careless making of unfair or unsustainable decisions adverse to the contractor.

The respective rights of the parties should be of such a nature that they might be fairly enforced whatever contingencies might arise and that, if such conditions were adopted, it should be understood by all the parties that in the event of a dispute arising every clause would be enforced without question. It might also be observed that the parties' obligations in their contractual arrangement have been based on the principle that the design of the permanent works is, generally, carried out by someone other than the contractor.

The central question which arises is that the structure of the contract into which the contractor was prepared to enter with the employer implies that the contractor will rely upon the engineer to execute properly the latter's duties.

In other words, although the parties were brought into close proximity in relation to the contract, a failure by the engineer or the architect to carry out their duties under the contract would foreseeably cause a loss to the contractor that was not properly recoverable under its rights against the employer.

Contra Proferentem Rule

The doctrine, as a general rule of construction, is stated in the Latin maxim *verba chartarum fortius accipiuntur contra proferentem.* As explained in Anson's Law of Contract, the rule is based on the principle that:

> a person is responsible for ambiguities in its own expression, and has no right to induce another to contract with it on the supposition that the words mean one thing, and then to argue for a construction by which they would mean another thing more to its advantage.

The *contra proferentem* doctrine is a deed or representation, to be construed more strongly against the party putting forward the document, whose purpose is to prevent the use of unintelligible terms through the threat of applying an interpretation in favour, not of whoever is responsible for creating such unintelligibility, but of the other party *ex ante* will (Kelleher and Walters 2020).

However, where both parties have been involved in agreeing the terms of a document, the courts will be reluctant to apply the *contra proferentem* doctrine; the doctrine only applies as a general rule of construction when all others have failed.

The *contra proferentem* rule is thus clearly one of default and is applied if, and only if, there is no particular condition that provides for the issue at stake; but it is also a rule which penalises the client or engineer's use of unintelligible industry common terms and conditions, as any doubts are always resolved against them and in favour of the contractor.

This leads to an outcome that is clearly opposed to what the author of the terms would have wanted, and thus acts as an incentive to word terms in a clear way, i.e. to reveal information to both the other party and the courts.

In summary, those drafting or amending contracts need to ensure that they incorporate terms that are concise, unambiguous, and, preferably, in plain and intelligible language. Otherwise, if there is doubt about the meaning of a written term, they may find that an interpretation more favourable to the other party is applied, in a particular circumstance. This may be contrary to their intentions, and, in certain circumstances, it could be financially disastrous (Beale et al. 2008).

Misrepresentation

Under contract law, a misrepresentation is defined as a false statement of a material fact, made by one party to another, which has induced the other party to enter into a contract. Even if the statement was one of opinion, an action for misrepresentation may also happen if it can be proved that the maker of the statement did not actually believe in the truth of the opinion or if it can be established that a reasonable man having the maker's knowledge could not honestly have held such an opinion (Allen and Martin 2010).

Misrepresentation is divided into the following categories:

- Innocent misrepresentation. This is a situation where the person making the statement can show he had reasonable grounds to believe his statement was true.
- Negligent misrepresentation. This describes a statement that is made carelessly or without reasonable grounds for believing its truth.
- Fraudulent misrepresentation. This occurs when a false statement is made knowingly, or without a belief that it is true, or with reckless disregard for its truth.

In respect to fraudulent misrepresentation, the following situations are tested by the courts:

1. There is no fraud if the person making the statement honestly believes the statement to be true.
2. A claimant will need to prove an absence of honest belief in the truth of the statement for an action for fraudulent misrepresentation to succeed.
3. It is enough to show that the person making the statement suspected it might be inaccurate, or that he neglected to make enquiries, without proving that the maker knew his statement was false.
4. Absence of reasonable grounds for a belief does not amount to fraud but may be used as evidence from which an inference can be drawn that there was no honest belief in the truth of the statement.

5. The test of misrepresentation is usually objective. Where the representation is claimed to be fraudulent, the court will inquire into the subjective state of mind of the maker of the statement.

The difficult question is whether, when an employer possesses relevant information, non-disclosure can amount to a misrepresentation. The answer depends on whether there is a duty to disclose the information. For example, insurance provisions in construction contracts are subject to a doctrine of utmost good faith, or *uberrimae fides*, which imposes a positive duty to disclose all material facts. There may be other situations where a party comes under an affirmative duty, but in relation to information on site and ground conditions, there is no general rule.

In addition, to be able to prove misrepresentation, the false statement must have induced the formation of the contract. However, the availability and effectiveness of the remedy is very limited. At common law, rescission would only be granted if it were possible to restore the parties precisely to their original positions.

The remedies afforded by the courts in case of misrepresentation are summarised thus:

1. Where a party has entered into a contract after a misrepresentation has been made to him by another party and as a result thereof he has suffered loss, then the party making the misrepresentation would be liable to damages, unless he proved that he had reasonable ground to believe and did believe up to the time the contract was made that the facts represented were true.
2. Where a party has entered into a contract after a misrepresentation has been made to him otherwise than fraudulently, and he would be entitled by reason of the misrepresentation to rescind the contract, then the court may declare the contract subsisting and award damages in lieu of rescission.
3. The misrepresentation needs not be the sole factor that induces the formation of the contract. Silence may amount to misrepresentation if a half truth is offered or if the maker realises the statement is not true before the contract is made. Rescission was considered an appropriate remedy because the false statement of fact could not be restored by payment of money (Kelleher and Walters 2020).

Mistake

To be operative, a mistake, in law, must be of fact and not of law. It allows the parties to rescind a contract, which effectively is to put the parties back into the positions they held before the contract was made.

The doctrine of mistake applies to facts present at the time of contract, while frustration applies to supervening events. This creates some confusion, as the courts have tended to treat the encountering of unforeseen ground conditions as a supervening event, although the conditions will normally have existed at the time of contract.

The term ‘mistake’ refers to situations where one or both parties to a contract are under a misapprehension of present fact at the time of contract. The situations can be divided for the purpose of legal analysis into two categories: unilateral mistake and common mistake (Furmston et al. 2006).

Unilateral mistake is where the mistake is such that the parties are at cross-purpose, or where the mistaken belief of one party is known to the other party. The doctrines relating to unilateral mistake are essentially extensions of the doctrines on the formation of contract, and the need for agreement.

Because of the lack of true agreement, the contract will be rescinded, and it is not difficult to say that the contract never really existed. There is a distinction between the various situations of unilateral mistake, as to the relevant test applicable. Where the mistake is known to one party, the test is subjective as to the belief of the mistaken person. When the parties are at cross purposes, the test is objective and relates to what a reasonable person would have understood from the agreement.

Common mistake, also called mutual mistake, is where the parties share the same misapprehension.

This includes cases where one party knows of the mistake but is unaware of, or does not consider, its significance. There is some disagreement about this terminology. Where the parties share a common misapprehension, the position is different as there is clearly agreement (Beale et al. 2008).

The law of common mistake sets out two conditions that are required for a contract to be held inoperative: firstly, the mistake should be sufficiently fundamental or basic to invalidate a contract, and secondly, the mistake must not be the responsibility of one or other of the parties.

Law cases show relief, granted in the case of a mistake in law, in four main situations:

1. In all situations, the mistake must be one of fact.
2. Where there has been a mistake as to the identity of the other party contracting and the first party did not intend to enter, and would not have entered, into a contract with that person.
3. A party signing a contract document has been misled as to the nature of the document, in which case the plea of *non est factum* may apply.
4. Where there is a clear mistake by the offeror as to the terms of the contract, and the mistake is known to the other party when he purports to accept.

Test of Reasonableness

The classical theory of a contract characterises it as a bargain resulting from an agreement between two equal parties. However, this characterisation overlooks the potential for this bargaining process to be distorted. There may be occasions on which a party, typically one with less 'power' (whether that be financial or otherwise), may agree to a term in the contract that places them at a disadvantage. Law has attempted to step in and reduce the likelihood of parties being unfairly prejudiced by such terms.

Reasonableness may be a measure of conduct but it is not a source of obligation. However, whether it is a standard of conduct, or an obligation flowing from the expectation of at least one and presumably both parties to a contract, it is necessary to consider how one determines a standard of reasonableness so that the parties to the contract and the courts may determine whether that standard has been adhered to. Notwithstanding an incapacity to define reasonableness, the courts have recognised its existence and importance in the law of contract (Beale et al. 2008).

The legal implication in contracts of what is reasonable runs throughout the whole of modern law in relation to business contract. Not being able to define the concept, but nonetheless recognising its existence and purporting to know what it means and how to apply it, the courts have sought to explain the concept by reference to other concepts, equally uncertain, such as fairness, good faith, or absence of unconscionable conduct. Furthermore, the explanation of reasonableness by reference to concepts such as fairness has occurred notwithstanding recognition that all commercial dealings between parties are not necessarily fair (Furmston et al. 2006).

Any examination of the concept of reasonableness quickly leads to the view that the standard is entirely subjective and probably indefinable. Judges have not sought to define reasonableness. Rather, discussion has focused upon the operation and application of the concept in particular circumstances. The law, in determining what is reasonable, is not concerned with ideal truth, but with something much less ambitious, though more practical.

Attempts to define reasonableness have led some to suggest that it is difficult, if not impossible, to ascribe any sensible meaning to the concept. Reasonableness, alone, is an abstract concept and does not by itself provide a test for determining what charges may or may not be made; it is a useful guide if, and only if, we are aware of the various matters which must be considered when the necessity arises of determining whether particular charges are or are not reasonable.

The concept of 'reasonableness' is used extensively in law and a number of tests have been developed to determine what constitutes 'reasonable'. Perhaps the widest use of the concept of reasonableness is found in the law of delict, which allows persons to claim for civil wrongs brought against them. One of the cornerstones of the law of delict is the test applied to the *diligens paterfamilias* or reasonable man. This test aims to establish whether a person acted negligently, and asks the following questions:

1. Would a reasonable person, in the same circumstances as the defendant, have foreseen the possibility of harm to the claimant?
2. Would a reasonable person have taken steps to guard against that possibility?
3. Did the defendant fail to take the steps that he or she should have reasonably have taken to guard against it?

If the above three questions are answered in the affirmative then negligence is established.

References

Books

Allen R, Martin S (2010) Construction law handbook. Aspen Publishers, MD
Beale H (2010) Chitty on contract. Sweet & Maxwell, London
Beale H, Bishop W, Furmston M (2008) Contract: cases and materials. Oxford University Press, Oxford
Beatson J, Burrows A, Cartwright A (2010) Anson's law of contract. Oxford University Press, Oxford
Campbell D (2019) Construction law in a nutshell, 2nd edn (Nutshell Series). West Academic, St. Paul
Furmston M, Cheshire GF, Fifoot CHS (2006) Cheshire, Fifoot and Furmston's law of contract. Oxford University Press, Oxford
Haidar AD (2011) Global claims in construction. Springer Verlag, London
Kelleher TJ, Walters GS (eds) (2020) Smith, Currie & Hancock's common sense construction law: a practical guide for the construction professional, 6th edn. Wiley, New York
McKendrick E (2009) Contract law. Palgrave Macmillan, Basingstoke, Hampshire
Richards P (2009) Law of contract. Pearson Longman, Harlow

Case Law

Carlill v Carbolic Smoke Ball Company [1892] EWCA Civ 1
Currie v Misa (1875) LR 10 Exch 153
Dunlop Pneumatic Tyre Co Ltd v Selfridge & Co Ltd [1915] AC 847
Tweddle v Atkinson (1861) 1 B&S 393

Chapter 3
Types of Construction Contracts

Abstract At the early stage of a project, the main issue that faces the employer and his consultants is to decide on the contract strategy that best suits the project objectives. The development of the contract strategy comprises a complete assessment of the choices available for the management of the design and the construction, to maximise the likelihood of achieving the project objectives. This chapter will review the types of construction contracts that are most commonly used, and for each type of construction contract a description, advantages and disadvantages are included to give guidance to contract managers, project managers, and other professionals in the construction industry. This chapter also deals with risks generally, including risks allocation, and contract provisions that both employer and contractor face during a project. The last section describes the principles that a contract manager needs to know when negotiating a contract, and the quantitative factors he must consider in choosing the appropriate contract for a particular project. The materials in this chapter are particularly relevant to all readers who need to have an understanding of project and construction management and contracts.

Keywords Build-operate-transfer (BOT) · Construction management · Cost plus · Design and build · Lump sum

Factors Affecting Contract Selection

The employer's priorities and objectives, and the type of works carried out, are the principle considerations affecting contract choice. The choice of procurement route and payment arrangements will also limit the contract options; hence, procurement and contract selection are considered in principle simultaneously.

The size and complexity of the contract vary accordingly. A proper contract selection for a project involves four key decisions:

1. setting the project objectives and constraints;
2. selecting a proper project delivery method;
3. selecting a proper contract form/type; and
4. contract administration practices.

A. D. Haidar, *Handbook of Contract Management in Construction*,
https://doi.org/10.1007/978-3-030-72265-4_3

As been mentioned in the abstract of this chapter, the main issue facing the employer, early on in a project, is to decide on the contract strategy that best suits the project objectives. The development of the contract strategy comprises a complete assessment of the choices available for the management of the design and construction to maximize the likelihood of achieving the project objectives.

In general, the following factors influence the choice of contract (Netscher 2017):

1. The nature of the employer.
2. The employer's risk attitude.
3. The employer's priorities.
4. The procurement method.

The Nature of the Employer

All construction works are ultimately undertaken for the benefit of the employer, who may be a private individual, an organization or a public body.

Public sector contracting authorities are typically experienced employers, and they are required to enter into fixed-price lump-sum contracts. Private employers, on the other hand, range from vastly experienced organisations to the totally inexperienced individual.

The needs of an experienced developer are very different to those of one-off employers, who may require considerable assistance in formulating their needs and understanding the nature and the responsibilities in a construction process.

Experienced and regular employers often have preferences for and expertise in specific procurement arrangements.

Inexperienced employers, on the other hand, need sound professional advice in choosing between contract options.

The Employer's Risk Attitude

This will depend to some extent on the nature of the employer and the works to be carried out.

Public employers tend to be more risk-averse than private employers, particularly those engaging in speculative developments, who may be prepared to take greater financial risks in order to achieve earlier completion.

Since the selected contract describes the agreement and allocates risks, the choice of contract should be the one offering the most value to the employer, bearing in mind their objectives.

The common employer requirements in a particular contract are:

1. Maximum price.
2. Fast contract start-up.
3. Cost control.
4. Detailed lump sum price.
5. Price certainty.
6. Value engineering.
7. Fast-track construction.
8. Design flexibility, accountability and control.
9. Design warranties.
10. Innovations.
11. Contractor insolvency (Patil and Woolhouse 2019).

The Employer's Priorities

The three most important considerations for any employer are usually cost, time and quality. The procurement process invariably calls for some compromise or consensus to balance these priorities. This requires high-level project definition, and financial know-how and risk analysis on the part of the employer.

Employers seeking cost certainty on a contract will favour, often, a lump sum arrangement. This requires enough time to complete the design or employer's requirements and increases the overall development time.

Incomplete design or insufficient performance requirements, however, lead to costly variations in both traditional and 'design and build' approaches.

Where time is the priority, fast-track management approaches using approximate quantities or cost plus arrangements may be favoured. None of these options provides cost certainty.

Where quality is the priority, the use of a 'design and build' contract will probably be inappropriate.

A further distinction exists between building and engineering contracts. In general, larger, more complex projects employ the emerging procurement routes and their associated detailed forms of contract. Traditional arrangements may not be suitable in these instances due to the need to resolve interface and clash issues among specialist contractors at the site level.

The Procurement Method

Many of the standard forms are drafted for use with particular procurement approaches, while others are flexible. Whether the design is the responsibility of the employer or the contractor is also a key determinant of contract choice.

Analysing the requirements and characteristics of the procurement system is vital, as employers generally fail to understand procurement methods practised in the industry.

The core objective of a procurement system is identifying responsibilities and tasks for each and every participant in a project.

Before looking at types of procurement, one should know the functions that are carried out in the construction industry and the parties involved with it. Generally, the following functions are undertaken in a project:

1. Design
2. Construct
3. Management
4. Finance
5. Hand-over.

Five procurement systems exist in the construction industry. Each of them will be discussed in detail in this chapter. They are:

1. Traditional Method.
2. Design and Build.
3. Direct Labour/Construction Management.
4. Joint Venture.
5. Build-Operate-Transfer (BOT)/ Private–Public Partnership (PPP).

Project Constraints

All construction projects have constraints that influence the achievement of the project objectives. These constraints should, therefore, be considered when choosing an appropriate contract strategy. There are a variety of constraints, and these are examples:

1. Availability of funds.
2. Availability of contractual incentives.
3. Method of tendering.
4. Project location.
5. Target dates of the project.
6. Possibility of design changes.
7. Availability of resources.
8. Applied working hours.
9. Number of contractors willing or able to tender.
10. Inflation (Netscher 2017).

Project Objectives

The employer will have a number of overall objectives. These objectives may be of primary and/or secondary importance.

Primary objectives include functional performance, time objectives and cost objectives, and can be summarised thus:

1. identify project criteria;
2. prepare scope of works including drawings, specifications and bill of quantities;
3. select contract type;
4. prepare tender documents;
5. ascertain the tenderers' prequalification and issue invitations;
6. organise tender meetings and provide a forum for questions and answers;
7. revise and evaluate the tenders; and
8. issue contract.

On the other hand, secondary objectives could arise on a construction project and would exert a major influence over contract strategy decisions. Examples of secondary objectives are:

- training of the employer's staff;
- transfer of technology;
- involvement of the contractor in design;
- involvement of the employer in contract management;
- choice of labour-incentive in construction;
- LEED;
- sustainability;
- HSE[1] requirements;
- QA/QC[2] methods;
- use of local material and resources; and
- protection of the environment.

Project Scope

Project scope encompasses the project deliverables and the necessary requirements, including expected work and output. This serves as an important guide for the execution of the project by the team members and provides the team with a detailed planning approach.

The project scope also defines the extent or the area that the contract covers. Any additions or omissions during the life of the project will increase or decrease the quantity of work involved. Likewise, any changes in the design must be analysed carefully to establish whether or not they are likely to affect the scope of the project.

[1] Health and Safety Executive.

[2] Quality Assurance and Quality Control.

Project scope includes three main elements, namely physical requirements, time and cost.

Requirements

Some of the requirements that are commonly used to identify project scope are[3]:

1. Earth works.
2. Civil/structural works.
3. Architectural works.
4. Mechanical works (heat ventilation, air conditioning, plumbing).
5. Electrical works.
6. QA/QC.
7. Permits and licenses requirements.
8. Bonds and insurances.
9. Safety.
10. Environmental and health.
11. Security.
12. Fire protection; Auxiliary.

Time

The scope of works and that of time are closely interrelated. Planning and properly scheduling a project are essential to identifying time constraints and managing time for completion, extension of time and how change orders affect time.

Decisions must often be made on the effect of increasing or reducing scope or time. If the completion date of a project is critical, then increasing its scope will call for an accelerated programme. The extra cost associated with this acceleration must be quantified.

Usually a detailed work plan for implementing the project scope will expound the timetable required for executing all activities, items and works in line with the overall project duration. The detailed work plan is typically submitted by the contractor within a period of 28 days from the date of project site handover.

The contractor shall explain in the aforementioned work plan the contract quantities and related cash flow, and the resources, which will be allocated through a specialised program (selected by the employer) such as Primavera, which shows all tasks and activities as well as the timescale.

The contractor shall implement all amendments requested by the employer or representative and provide periodic and detailed updates in the manner and at the time specified by the employer or engineer. Upon approval of the time schedule by the

[3]They differ from the employer's requirements, as identified previously in this book.

supervisory body, the duration for carrying out the specified work shall determine the timescale for work completion provided in the detailed work plan (Netscher 2017).

Cost

The cost of a project is closely related to scope and time. The effect of the contract on price, and the various incentives and penalties that can help to keep the price steady, must be discussed and clearly defined.

Costing is a continuous process. It starts with the bidding process and is continued throughout the project through value engineering, procurement process and invoicing.

The cost is itemised in a bill of quantities and schedule of rates depending on the type of contract chosen.

The cost of a project includes, generally:

1. Indirect costs. These are the main office and site management overheads.
2. Direct costs. These are the labour costs.
3. Materials and equipment costs.

Project Delivery Methods

The project delivery method determines the project parties who are involved in the project and how they interact with each other. This is also called 'project organizational structure'.

The choice of an organizational structure should be related to the project objectives and constraints. It can be facilitated with consideration of the following factors:

1. Size and nature of the work packages within the project.
2. Selection of the design team, in-house resources, external consultants and contractors.
3. The process of supervision of construction.
4. Restrictions on using a combination of organizational structures within the project.
5. Expertise which the employer wishes to commit to the project.

When the design is at the development stage and the employer is interested in securing a low bid, the use of competitive bids is suggested. Competitive bidding results in a type of contract that many are familiar with.

A negotiated contract is used when the construction process starts before plans are completed, or when the many unknown factors of a project make an accurate estimate impossible. When many changes are expected and when inspection and supervision cannot be done efficiently, the negotiated type of contract should be used (Netscher 2017).

The various project delivery methods are consistent with the procurement methods and are discussed in more detail herein.

Contract and Procurement Arrangements

Depending on the type of works, the employer's identity and requisites and the type of financing, there are a multitude of contracts and procurement arrangements that can be used (Chappell 2020).

Traditional Approach

This is the most common approach in civil engineering projects; one in which the design has to be completed before construction can start. Design and construction are usually performed by two different parties who interact directly and separately with the employer.

This arrangement typically involves the employer appointing consultants to produce a design, select a contractor and supervise the works up to completion. The contractor is usually selected on the basis of a bidding process and is responsible for the management and delivery of the project and for the quality of workmanship and materials used, including those of the subcontractors.

This process has been extensively used in the past and continues to be widely used. Traditional procurement is suited for a wide range of projects including those where the scope of the works contain limited unknowns and risks, which the contractor would be unable to establish in advance. The approach, however, may not be suited to particularly complex projects requiring advanced management systems, structures and skills.

Developers and employers who can avail themselves of in-house design and management services, or who wish to repeat a pre-existing standard design, may favour this approach.

The traditional procurement approach is associated with projects where the employer prioritises the aesthetic and quality aspects of the project. It allows him to maintain control over the quality of the works during the design and construction phases of the project.

The process is typically standard in approach and proceeds through a succession of design development stages that allow sufficient time for the design teams to explore options and to finalise the works requirements within budgetary and time constraints.

This process allows comprehensive tender documentation to be developed, which facilitates competitive tendering arrangements, and enables lump sum contracts to be agreed with contractors. The outcome will, therefore, produce cost certainty.

Nevertheless, costs may need to be subsequently adjusted, as the arrangement provides the flexibility for employers to appoint named or nominated subcontractors to carry out specialist aspects of the design, and the ability to introduce variations and scope changes to cater for contingencies and the changing needs of the employer (Patil and Woolhouse 2019).

The main drawbacks of the traditional approach derive from the separation of the design from the construction function. The technical and management personnel involved in the project are not part of a single integrated team with sole responsibility for design and construction.

Contractors, who are construction experts, are largely excluded from the design development process, with the result that more buildable, innovative, or sustainable solutions may have been overlooked.

This separation also complicates communications and may foster an 'us and them' attitude which reduces the team spirit vital for successful projects. This difficulty is compounded by the competitive tendering process associated with traditional procurement, which may lead to fractious relationships during the delivery of a project.

The separation also leads to difficulties in establishing liability for defects as it may be unclear whether these arise from design or construction shortcomings, or a combination of both.

The process is viewed as slow, as full design is required before tenders can be obtained, thereby incurring additional financing costs for the employer.

Where a full design is not provided at the tender stage, additional information will be required during the construction phase, often resulting in variations, delays, and disputes.

It is often noticed that traditional projects are, all too frequently, late, over budget, and prone to disputes.

The main advantages of the traditional method are:

- Price competition,
- Total cost is known before construction starts,
- Well-documented approach, used in most government projects.

The main disadvantages of the traditional method are:

- Long time to complete.
- Design does not benefit from construction expertise.
- Conflict between employer, contractor and architect/engineer.

Therefore, this method is acceptable in many cases where the project is clearly definable, design is completed, time need not be shortened and changes are unlikely to occur during construction.

Construction Management: Direct Labour

The Direct Labour method, also called Management Contracting or Construction Management, is a method where the overall design is the responsibility of the employer, and the contractor is responsible both for defining packages of work and then for managing the carrying out of the work through separate trade or works contracts. In management contracting works, as opposed to construction management, contractors are employed by the management of the contractor, which is typically a leading construction firm (Chappell 2020).

Management approaches are associated with high-end, large-scale, complex and fast-moving projects where the employer seeks early completion prompted by the costs associated with building the project.

The main contractor or construction manager is usually highly experienced and reputable and is paid a management fee to contribute expert advice on constructability, programming and coordination issues to manage the execution of the work.

The appointment of the construction manager is made at an early stage in the design development process and the contractor/consultant works alongside the design team members. The works are constructed by specialist subcontractors, who can either be obtained by competitive tender or nominated by the employer.

In construction management the services may be provided by an individual or a firm acting as an agent for the employer, and the workers and subcontractors are directly engaged by the employer. In effect there is no main contractor.

This method is characterised by the following:

- It is used by large authorities.
- The employer performs both the design and the construction.
- Consultants are employed for some specialized designs.
- It is most suitable for small projects.
- It can be used when expertise is available.
- It may suit low-risk projects.
- It often entails inadequate scope definition.

The main advantages of construction and contracting management using direct labour are:

- Management approaches capitalise on the contractor's early involvement, which helps to produce the practical and buildable design.
- The approach is fast track; overlapping site operations and detailed design work can proceed in parallel, facilitating an early start on site.
- This benefit is compounded by the manager's expertise in coordination, programming and control, which fosters greater operational speed and efficiency.
- The appointment of the contractor within the design team or as the employer's agent removes the tensions associated with the usual objectives of maximising profit through pursuing variations and generating claims during the construction stage.

- The contractor/manager is paid a fee and looks after the employer's interests. This often results in more harmonious relationships during the project.
- The employer bears the bulk of the commercial risk under management procurement arrangements.

Some of the disadvantages are:

- There is no firm contract price before the work starts on-site, and the decision to proceed usually has to be taken on the basis of a cost plan and cost certainty that may not be achievable until much of the construction operations have been completed.
- The employer must have sufficient knowledge and flexibility to accommodate the evolving design decisions and contingencies as the works progress.
- The combination of superior quality and fast delivery requires a professional procurement process and a dependable supply chain.
- The approach increases design team consultancy fees and tends to use reliable work packages by contractors with proven credentials. These are usually more expensive, despite the use of competitive tendering.
- Management approaches also suffer from unclear lines of liability in the event of a project failure, and there may be difficulties in establishing whether the manager or the relevant package contractor is liable (Patil and Woolhouse 2019).

Design and Build (D&B)

Under this arrangement, the contractor undertakes both the design and the construction of the work in return for a lump sum price.

The contractor may be appointed either by competitive tender or as a result of a negotiated agreement. Although the design and build arrangements can be employed on sophisticated buildings, they tend to be associated with projects where the employer is more concerned with the functional aspects of the building rather than its aesthetic qualities. Hotels, large housing units and office buildings are typical examples of this approach.

Design and build is appropriate for new construction where the employer's requirements can be clearly defined. It is less successful on projects where planning or scope risks may mean that the employer's requirements cannot be comprehensively defined in advance of appointing a contractor.

The specifications play a crucial role in a design and build project. Moreover, detailed and up-to-date specifications can lessen the risks of future variations and claims.

There are a number of benefits claimed for the design and build arrangement, and these can be summarised as follows:

- The approach is characterised by an integrated design and construction structure, which eliminates many of the problems associated with traditional procurement arrangements.

- It seeks to establish sole responsibility for the design and delivery of the project within a single organisation.
- Relationships and communications are more straightforward, as is establishing liability if things go wrong.
- Design and build arrangements are 'fast track', allowing for the overlapping of design and construction activities. This characteristic allows construction work to start early while much of the downstream detail design work is developed in parallel with site operations, thereby enabling earlier completion dates to be achieved.
- Producing competitive design requires an engagement in value engineering processes, particularly where contractors have ongoing obligations during the operation of the facility. This may result in more buildable and sustainable solutions being developed.
- The arrangement is lump sum, based on the contractor's proposals, promising greater cost and time certainty as all design proposals and information flow requirements are addressed within the contractor's organisation.
- More risks are transferable to the contractor, who is responsible for design.
- It involves minimum employer involvement (Harris et al. 2013).

Some of the drawbacks of a design and build approach are:

- The success of the design and build arrangement largely hinges on the ability of the employer to produce a definitive project brief; a process that usually requires the appointment of consultants. This process takes time; incomplete briefing information will result in inadequate proposals being developed.
- Correcting these inadequate proposals by introducing variations may cause significant disruption in the contractor production process, and it is often difficult to ascertain their cost implications.
- The contract sum analysis supporting the contractor proposals is typically based on preliminary design[4] and is, therefore, not quantifiable to the level of detail associated with traditional approaches.
- The individual elements of the descriptions contained in the contract sum analysis, for example, design, labour, overheads and profit, are not usually identifiable.
- The valuation of variations is the subject of negotiated 'fair' rates,[5] and often costs are significantly more than may be anticipated under traditional arrangements. The arrangement is regarded as less flexible as a result.
- Cost may not be known until the end of the construction.
- Design and build contactors may reduce quality to save cost.

The use of this approach, therefore, should be considered when contractors offer specialised design, construction, or other expertise, or when the design is strongly influenced by the method of construction.

[4]Or tender design.

[5]The inclusion of a schedule of rates is important in D&B projects, as it can establish some unit rates for calculating a variation.

Build-Operate-Transfer (BOT)

In this approach, a business entity is responsible for performing the design, construction, long-term financing, and temporary operation of the project. At the end of the operation period, which is usually more than ten years, the operation of the project is transferred to the employer.

This approach has been extensively used in recent years and is expected to continue. An example of its use is express routes, turnpikes, electrical stations, desalination plants and sewage plants.

In a BOT project a consortium of companies shares the cost (design, construction, financing, operation, and maintenance) and the profits gained from user fees, for a stipulated number of years. Afterwards, the project returns to the government to become publicly owned.

This approach has also been used extensively in large infrastructure projects financed by the World Bank in parts of the world that cannot afford the high investment cost of such projects.

Advantages of Build-Operate-Transfer

1. It minimizes the public cost for infrastructure development.
2. It reduces public debt.
3. It allows for innovation.
4. It provides a chance to bring in expertise.
5. It allows each party to focus on their strengths.
6. It keeps public-sector funds where they are most needed.

Disadvantages of Build-Operate-Transfer

1. It can have higher transaction costs.
2. It only works for large projects.
3. It requires fundraising to be successful.
4. It requires substantial operational revenues to be successful.
5. It requires strong corporate governance.
6. It can place the public sector at a disadvantage.

PPP—Public–Private Partnership

There is no standard definition of a PPP model. It is not uncommon for PPP to refer to a wide spectrum of procurement models, to describe private sector participation in public service delivery.

The PPP approach must fulfil three main criteria, which are:

1. The project is an investment project belonging to the state and involving an undertaking in respect of which a particular state agency has duties and powers to carry out under the law.

2. It either authorises a private party to invest by way of permission, concession or licensing in any form whatsoever, or else it enters into partnership with a private party.
3. It is related to infrastructure and public services as specified under government laws, such as roads, highways, expressways, railways, mass transit or airports.

PPPs are a viable avenue for governments to benefit from private capital and technical know-how.

Both public and private sector stakeholders should understand the interaction of factors that make a PPP project commercially viable, given the varying levels of public sector capabilities, PPP frameworks and economic potential.

A structured PPP project would have a well-balanced allocation of risks and returns between stakeholders in the public and private sector.

Detailed feasibility studies are required to determine the project's commercial viability and risk allocation. This process, though reassuring to private sector bidders, can be costly and time-consuming (Harris et al. 2013).

Generally, PPP contracts must at least contain the following clauses:

1. Duration, provision of services and implementation of the project.
2. Rights and duties of each party.
3. The ownership of the project assets and their valuation.
4. Changes to the nature of the provision of services under the project.
5. Changes to a contracting party, contractor or subcontractor and the assignment of claims.
6. Force majeure (Acts of God).
7. Termination of the contract.
8. Step-in rights and details thereof.
9. Dispute resolution.

Additional requirements must be implemented in a PPP contract and are as follows:

1. Background, objective and scope of project.
2. Definition and interpretation.
3. Hierarchy of documents.
4. Source of funds and investments.
5. Operation, output specification and level of service.
6. Transfer of knowledge.
7. Returns each party is entitled to, and ways to provide such returns.
8. Public benefit.
9. Governance and monitoring of project operation.
10. Tax, fees, interest, and bank fees.
11. Warranty.
12. Insurance.
13. Changes in laws.
14. Security and guarantees.

Types of Construction Contracts

While construction contracts serve as a means of pricing construction, they also structure the allocation of risks to the various parties involved.

The employer has the sole power to decide what type of contract should be used for a specific facility to be constructed and to set forth the terms in a contractual agreement. Hence, it is important to understand the risks of the contractors associated with different types of construction contracts (Netscher 2017).

Lump Sum Contract

A lump sum contract is normally used in the construction industry to reduce design and contract administration costs. It is called a lump sum because the contractor is required to submit a total and global price instead of bidding on individual items.

A lump sum contract is the most recognized agreement form and it is usually used in simple and small projects, projects with a well-defined scope, and construction projects where the risks associated with different site conditions are minimal.

A lump sum contract is an efficient contract agreement to be used if the requested works are well defined and construction drawings are completed. The lump sum agreement will reduce the employer risk, and the contractor has greater control over profit expectations.

In a lump sum contract, the employer, essentially, assigns all the risks to the contractor, who in turn can be expected to ask for a higher mark-up in order to take care of unforeseen contingencies. A contractor being contracted under a lump sum agreement will be responsible for the proper execution of the job, and will provide its own means and methods to complete the work.

This type of contract is usually developed by estimating labour costs and material costs and adding a specific amount that will cover the contractor's overhead and profit margin.

The amount of overhead calculated under a lump sum contract will vary from one contractor to the next. It will be based on their risk assessment study and general expertise. Estimating a very large overhead cost, however, can lead the contractor to present higher construction costs to the employer.

The expertise of the contractor will determine their estimated profit. A poorly executed and long-delayed job will raise construction costs and eventually diminish the contractor's profit.

It is also a preferred choice when stable soil conditions, full pre-construction studies, and assessments are completed and the contractor has analysed those documents.

The stipulated sum contract might contain, when agreed upon by the different parties, certain unit prices for items with indefinite quantities and allowance to cover any unexpected conditions.

The time to award this type of contract is also longer; however, it will minimise change orders during construction.

A lump-sum contract offers the following advantages:

- There is a low risk to the employer.
- There is a fixed construction cost.
- It minimizes change orders.
- Employer supervision is reduced when compared to 'Time and Material Contract'.
- The contractor will try to complete the project faster.
- It is accepted widely as a contracting method.
- Bidding analysis and selection process is relatively easy.
- The contractor will maximize its production and performance (Harris et al. 2013).

Although lump sum contracts are the standard and preferred option for most contractors, it might also have some limitations:

- *Unbalanced Bids* Some projects might require an application for payment using unit quantities and unit prices. Many contractors will produce an unbalanced bid by rising unit prices on items to be completed early in the project, such as mobilisation, excavation and structural works, and lowering unit prices on items needed in later stages.
- *Change Orders* If the employer produces or receives a change order proposal from the contractor, the price quotation could possibly be disputed. The employer might appeal that the requested change was already covered under contract provisions. It is important to prepare specific contract clauses specifying how change orders are going to be managed and to what extent the contractor could claim delay damages.
- *Scope and Design Changes* A contractor may suggest design changes based on their experience. Contract provisions should be clear on how those changes will be addressed and how the costs will be divided, or who will be responsible for the economic impact of the proposed changes.
- *Early Completion* Lump-sum contracts might include early completion compensation for the contractor. Early completion might produce higher savings for the employer; however, those clauses must be explicit in the construction contract.

In a lump sum contract, the employer essentially assigns all the risks to the contractor, who in turn can be expected to ask for a higher mark-up in order to take care of unforeseen contingencies. Besides the fixed lump sum price, other commitments are often made by the contractor in the form of submittals such as a specific schedule, the management reporting system or a quality control programme.

If the actual cost of the project is underestimated, the underestimated cost will reduce the contractor's profit by that amount. An overestimate has an opposite effect, but may reduce the chance of being a low bidder for the project (Eggink 2020).

Unit Price Contract

The unit pricing contract is another type of contract commonly used by government agencies or large institutions carrying out identical works of defined scope but different size.

Unit prices can also be set during the bidding process of a lump sum contract or re-measured contract as the employer requests specific quantities and pricing for a predetermined amount of some items that might be added to the works. (A design-build project is a typical kind of contract where unit prices are included as a separate supplement.)

By providing unit prices, the employer can verify that he has been charged with uninflated prices for labour, materials and services being acquired.

Unit price can easily be adjusted up and/or down during scope changes, making it easier for the employer and the contractor to reach an agreement during change orders.

This type of contract is used for routine contracted service requirements, where the total value of the contract can be calculated by multiplying identical units of work by a fixed unit price. Generally, a unit price contract will not be used to cover a wide range of services of a similar nature but may be appropriate when a specific service is required for a defined period of time. In such cases, the total amount of work may not be known.

They may also be appropriate for reoccurring contracts that are carried out on a routine basis, such as maintenance (boilers, ACs,[6] paint jobs etc.) or road repairs.

A unit price contract must include clear provisions such as:

1. The scope of the intended works requirements should be clearly defined.
2. The establishment of unit price contracts should be subsequent to the competitive bidding process unless the requirement meets the criteria for sole source contracts.
3. There should be the inclusion of assessments, minor shop materials, tools, tackle, local transportation, overhead, profit, and any charges associated with providing the service after normal business hours.
4. The risk of inaccurate estimates for uncertain quantities for some key tasks shall be removed from the contractor offer.

While the main advantage of unit price contracts is that they are well-suited for repetitive and well-defined tasks, they are not particularly useful for most private building projects, except as part of a lump sum or cost plus contract, applied to select components of work items such as dirt removal or fill, finish hardware, etc.

[6] Air conditioning units.

Cost Plus Contract

Cost plus contracts are often used when the scope has not been clearly defined, such as when the project design is still being finalized and the employer wants to begin construction. An employer agrees to pay the cost of the work, including all trade subcontractor work, labour, materials, and equipment, plus an amount for the contractor's overhead and profit (Eggink 2020).

Since the contractor is reimbursed only for actual costs, plus a fee for overhead and profit, if actual costs are lower than estimated, the employer gets to keep the savings. If the actual costs are higher than estimated, the employer must pay the additional amount, unless the cost is capped at a guaranteed maximum price.

The advantage of a cost plus contract is that the project will result in what was intended, even if costs run high. However, despite the lower amount of risk for the contractor, these contracts are harder to track and more supervision is needed. The most common cost plus contracts are:

- Cost plus fixed fee.
- Cost plus fixed percentage contract.
- Cost plus—with a guaranteed maximum price.
- Cost plus variable percentage contract.
- Target estimate contract.

Cost Plus Fixed Fee

Cost plus fixed fee is a pricing option in which a party providing the work calculates the cost to complete the work, and then adds a fee.

Costs to complete the work typically include all amounts paid to subcontractors, labour and materials, plus overhead—or those items that are outside direct labour and materials, such as project management, site office, tools and equipment.

The fee, which is the profit estimated by the contractor, and with some contracts all or part of the overhead, may be one defined amount or it may be based on a percentage of the defined costs.

In a standard cost plus fee contract, the additional fee is not intended to be calculated as a percentage measure of the total costs, in which the fee in these situations would vary based on the actual costs. Instead, the cost plus fixed fee contract provides for a predetermined fixed fee reimbursement. Cost plus fixed fee tends to be more advantageous to the employer, as opposed to the contractor, as it caps the fee and the fee will not change based on the future expansion or fluctuations of the budget of the project.

Cost Plus Fixed Percentage Contract

For certain types of construction involving new technology or extremely pressing needs, the employer is sometimes forced to assume all risks of cost overruns. The contractor will receive the actual direct job cost plus a fixed percentage, and have little incentive to reduce job cost. Furthermore, if there are pressing needs to complete the project, overtime payments to workers are common and will further increase the job cost.

Unless there are compelling reasons, such as urgency in the construction of military installations, the employer should not use this type of contract.

This type of contract involves payment of the actual costs, purchases or other expenses generated directly from the construction activity. Cost plus contracts must contain specific information about a certain pre-negotiated amount (some percentage of the material and labour cost) covering contractor's overhead and profit.

Cost plus contracts are used when the scope has not been clearly defined and it is the employer's responsibility to establish some limits on how much the contractor will be billed. When some of the aforementioned options are used, those incentives will serve to protect the employer's interest and avoid him being charged for unnecessary changes.

Cost Plus—with a Guaranteed Maximum Price

It is not hard to imagine a pricing method based on costs, which are often a moving target, escalating out of control. That is why cost plus pricing is often combined with a guaranteed maximum price (GMP). Cost plus with GMP provides an upper limit on total construction costs and fees for which an employer is responsible. If the party providing the work under this pricing method runs over GMP, it is responsible for such overruns. The parties to a cost plus with GMP contract negotiate at the outset how any costs savings (i.e. work performed for less than GMP) will be allocated, often agreeing to share in these savings.

Advantages of a cost plus with a guaranteed maximum price contract are mainly:

1. For projects that need to be fast-tracked, cost plus pricing enables a contractor to begin work earlier with preliminary phases of construction while design of the project is being completed.
2. Cost plus with GMP and an agreement for sharing cost savings can incentivize both parties to a construction contract to work together as efficiently as possible.

The main disadvantages are:

1. Cost plus without GMP leaves an employer responsible for paying for workers exposed to unrestrained liability for costs.
2. Cost plus pricing requires more work from both parties to the contract; the contractor providing the work must track and report costs, and the employer obligated to pay for the work must analyse this data for accuracy.

Cost Plus Variable Percentage Contract

For this type of contract, the contractor agrees to a penalty if the actual cost exceeds the estimated job cost, or a reward if the actual cost is below the estimated job cost.

In return for taking the risk on its own estimate, the contractor is allowed a variable percentage of the direct job cost for its fee. Furthermore, the project duration is usually specified and the contractor must abide by the deadline for completion.

This type of contract allocates considerable risk for cost overruns to the employer, but also provides incentives to contractors to reduce costs as much as possible.

Target Estimate Contract

This is another form of contract which specifies a penalty or reward to a contractor, depending on whether the actual cost is greater than or less than the contractor's estimated direct job cost. Usually, the percentages of savings or overrun to be shared by the employer and the contractor are predetermined and the project duration is specified in the contract. Bonuses or penalties may be stipulated for different project completion dates.

Risks Associated with the Contractor–Employer Relationship

The key influencing risk factors that affect the relationship between the parties to a contract are:

1. Trust (mutual trust or suspicion/mistrust).
2. Objectives (common or self-objectives).
3. Teamwork or fragmentation.
4. Risk allocation (sharing risks or not).
5. Continuous improvement, or a lack of it.

6. Communication (open and effective, or ineffective).
7. Business attitude (win–win or win–lose).
8. Problem solving/conflict resolution.
9. Procurement/competitive tendering/contract.
10. Senior management commitment.
11. Sharing information and learning, or withholding information.
12. Focus (long-term or short-term).
13. Flexibility to change, or resistance to change.
14. Lack of partnering experience.
15. Incentives.
16. Performance assessment.
17. Transparency.
18. Monitoring.

Risks associated with the project works, tender, design and procurement are:

1. Change in scope of work.
2. Insufficient design completion during tender invitation.
3. Unforeseeable design development risks at tender stage.
4. Errors and omissions in tender document.
5. Exchange rate variations.
6. Unforeseeable ground conditions.
7. Actual quantities of work required far exceeding estimate.
8. Lack of experience of contracting parties.
9. Inflation beyond expectation.
10. Unrealistic maximum price or target cost agreed in the contract.
11. Disagreement over evaluating the revised contract price after submitting an alternative design by main contractor.
12. Difficulty of the main contractor to have back-to-back contract terms with nominated or domestic subcontractors.
13. Global financial crisis.
14. Poor buildability or constructability of project design.
15. Delay in resolving contractual disputes.
16. Loss incurred by the main contractor due to unclear scope of work.
17. Delay in work due to the third party.
18. Inclement weather.
19. Inaccurate topographical data at tender stage.
20. Little involvement of main contractor in design development process.
21. Selection of subcontractors with unsatisfactory performance.
22. Impact of construction project on surrounding environment.
23. Technical complexity and design innovations requiring new construction methods and materials from main contractor.
24. Market risk due to the mismatch of prevailing demand.
25. Change in interest rate on main contractor's working capital.
26. Environmental hazards of constructed facilities.
27. Delay in availability of labour, materials and equipment.

28. Delayed payment.
29. Change in relevant government regulations.
30. Force majeure.

Contract Provisions for Risk Allocation

Provisions for the allocation of risk among parties to a contract can appear in numerous areas in addition to the total construction price. Typically, these provisions assign responsibility for covering the costs of possible or unforeseen occurrences (Eggink 2020).

A partial list of responsibilities with associated risk that can be assigned to different parties would include:

- Force majeure (this provision absolves an employer or a contractor for payment for costs due to 'Acts of God' and other external events such as war or labour strikes).
- Indemnification (this provision absolves the indemnified party from any payment for losses and damages incurred by a third party such as adjacent property employers).
- Liens (assurances that third-party claims are settled such as 'mechanics liens' for workers' wages).
- Labour laws (payments for any violation of labour laws and regulations on the job site).
- Differing site conditions (responsibility for extra costs due to unexpected site conditions).
- Delays and extensions of time.
- Liquidated damages (payments for any facility defects with payment amounts agreed to in advance).
- Consequential damages (payments for actual damage costs assessed upon impact of facility defects).
- Occupational safety and health of workers.
- Permits, licenses, laws, and regulations.
- Equal employment opportunity regulations.
- Termination for default by contractor.
- Suspension of work.
- Warranties and guarantees.

Principles of Contract Negotiation

Negotiation is another important mechanism for arranging construction contracts. Project managers often find themselves as participants in negotiations, either as principal negotiators or as expert advisors.

These negotiations can be complex and often present important opportunities and risks for the various parties involved.

For example, negotiation on work contracts can involve issues such as completion date, arbitration procedures, special work item compensation, and contingency allowances, as well as the overall price.

As a general rule, factors such as the history of a contractor and the general economic climate in the construction industry will determine the results of negotiations. The skill of a negotiator can, however, affect the possibility of reaching an agreement, the profitability of the project, the scope of any eventual disputes and the possibility for additional work among the participants (Godwin 2012).

Thus, negotiations are an important task for many project managers. Even after a contract is awarded on the basis of competitive bidding, there are many occasions in which subsequent negotiations are required as conditions change over time.

In conducting negotiations between two parties, each side will have a series of objectives and constraints. The overall objective of each party is to obtain the most favourable, acceptable agreement.

Poor negotiating strategies adopted by one party or another may also preclude an agreement even with the existence of a feasible agreement range. For example, one party may be so demanding that the other party simply breaks off negotiations. In effect, negotiations are not a civilised solution methodology for the resolution of disputes.

In light of these tactical problems, it is often beneficial to all parties to adopt objective standards in determining appropriate contract provisions. These standards would prescribe a particular agreement or a method to arrive at appropriate values in a negotiation. Objective standards can be derived from numerous sources, including market values, precedent, professional standards, what a court would decide, etc. By using objective criteria of this sort, personalities and disruptive negotiating tactics do not become impediments to reaching mutually beneficial agreements.

With additional issues, negotiations become more complex both in procedure and in the result. With respect to procedure, the sequence in which issues are defined or considered can be very important. For example, negotiations may proceed on an issue-by-issue basis, and the outcome may depend upon the exact sequence of issues considered. Alternatively, the parties may proceed by proposing complete agreement packages and then proceed to compare packages (Godwin 2012).

With respect to outcomes, the possibility of the parties having different valuations or weights on particular issues arises. In this circumstance, it is possible to trade off the outcomes on different issues to the benefit of both parties. By yielding on an issue of low value to oneself but of high value to the other party, one may obtain concessions on other issues.

To focus the negotiations, certain issues are predefined in a contract:

- *Duration* The final contract must specify a required completion date.
- *Penalty for Late Completion* The final contract may include a daily penalty for late project completion on the part of the contractor.

- *Award for Early Completion* The final contract may include a daily bonus for early project completion.
- *Frequency of Progress Reports* Progress reports can be required daily, weekly, bi-weekly or monthly.
- *Contract Type* The construction contract may be a flat fee, a cost plus a percentage profit, or a guaranteed maximum with cost plus a percentage profit below the maximum.

Relative Costs of Construction Contracts

Regardless of the type of construction contract selected by the employer, the contractor recognizes that the actual construction cost will never be identical to the contractor's own estimate because of imperfect information and external conditions not accounted for.

Furthermore, it is common for the employer to place work change orders to modify the original scope of work, for which the contractor will receive additional payments as stipulated in the contract.

The contractor will use different mark-ups commensurate with its market circumstances and with the risks involved in different types of contracts, leading to different contract prices at the time of bidding or negotiation.

The type of contract agreed upon may also provide the contractor with greater incentives to try to reduce costs as much as possible.

Therefore, the contractor's gross profit on completion of a project is affected by the type of contract, the accuracy of its original estimate, and the nature of work change orders.

The employer's actual payment mechanism for the project is also affected by the contract and the nature of work change orders.

In order to illustrate the relative costs of several types of construction contracts, the pricing mechanisms for such construction contracts must be formulated in the same direct job cost and must include corresponding mark-ups, and other factors as follows:

1. Contractor's original estimate of the direct job cost at the time of contract award.
2. Amount of mark-up by the contractor in the contract.
3. Estimated construction price at the time of signing contract.
4. Contractor's actual cost for the original scope of work in the contract.
5. Underestimate of the cost of work in the original estimate.
6. Additional cost of work due to change orders.
7. Actual payment to contractor by the employer.
8. Contractor's gross profit.
9. Basic percentage mark-up above the original estimate for fixed-fee contract.
10. Premium percentage mark-up for contract type.

11. A factor in the target estimate for sharing the savings in cost as agreed upon by the employer and the contractor (Godwin 2012).

References

Chappell D (2020) Construction contracts, 4th edn. CRC Press

Eggink E (2020) A practical guide to engineering, procurement and construction contracts. Taylor and Francis, Abingdon

Godwin W (2012) International construction contracts. Wiley-Blackwell, Hoboken, NJ

Harris F, McCaffer R, Edum-Fotwe F (2013) Modern construction management, 7th edn. Wiley-Blackwell, Hoboken, NJ

Netscher P (2017) Construction management: from project concept to completion. Panet, Subiaco

Patil BS, Woolhouse SP (2019) BS Patil's building and engineering contracts, 7th edn. CRC Press, Boca Raton

Chapter 4
Standard Forms of Contract

Abstract This chapter will look at the main standard forms of contract that are being used in the USA, the UK and in other countries, focusing, of course, on the most popular form—the FIDIC contract. The purpose of standard forms of contract is to facilitate the contractual arrangements between parties in a construction project. Standard forms of contract offer ready-made terms and conditions when one is making a contract. These standards are commonplace in construction transactions and generally accepted by the different contracting parties. The literature on standard form contracts in construction has increased dramatically in recent years as they are becoming more and more popular due to their ease of use and the fact that employers and contractors are becoming familiar with their use. They also provide a good platform to start a contract and are written so that changes can easily be made to them in the form of special (particular) conditions without changing the general terms and conditions.

Keywords AIA (American Institute of Architects) · DBIA (Design-Build Institute of America) · ICE (Institution of Civil Engineers) conditions of contracts · NEC (New Engineering Contracts) · JCT (Joint Contracts Tribunal) · FIDIC (international projects mostly use forms available from the International Federation of Consulting Engineers)

Introduction

Standard forms of contract account for the bulk of contracts made in business agreements and construction contracts.

The standard clauses in these contracts have been settled over the years through negotiation by clients, engineers and lawyers and have been widely adopted because experience has shown that they facilitate the commercial and contractual conduct between the parties.[1]

---

[1]Contracts of these kinds affect not only the actual parties that are party to them, but also others who may have an interest in the transactions to which they relate.

A. D. Haidar, *Handbook of Contract Management in Construction*,
https://doi.org/10.1007/978-3-030-72265-4_4

The fact that they are widely used by parties whose bargaining powers are fairly matched raises a strong presumption that their terms are fair and reasonable.

Often standard forms of contracts are of use because the parties regularly enter complex technical and legal relations. This is the case, for instance, in the construction and engineering industries, because the dealings in question involve transactions relating to projects of complexity and value. The basic reason underlying the widespread use of standard forms of contract in construction is the need to facilitate the conduct of operations and works, including the interrelationships and liabilities of the parties involved in a project, in the most efficient way (Close 2017).

A major reason for the development of standard written contracts is that they are valuable to the construction industry in particular. The JCT forms of building contract and, particularly, FIDIC suites of contract, discussed in this chapter, are good examples of this. With these types of contracts, there is a presumption by the courts that they are fair and reasonable. This does not mean, however, that standard forms of contract are without their problems (Godwin 2012).

The most cited reasons for using standard forms, and the reasons for their long-standing popularity, include:

- Reduced drafting time.
- Provision of a checklist of items to be agreed between the parties.
- Provision of a negotiation benchmark.
- Benefit from case law on the interpretation of terms and/or the impact of legislative change.
- The benefit of familiarity.

The prevalence of standard form contracts is such that the courts are regularly asked to interpret them. A dispute between the parties to a standard form of contract may, for instance, require the court to establish the true meaning of an individual clause, the relationship between two printed clauses, or the standing of a printed clause and a written addition in the form of a particular condition.[2] It may involve a question of whether or not a term may be implied in the contract, or the breadth of an exclusion or limitation of a liability clause (Klee 2017).

One of the main attractions of using standard forms is the fact that they are produced by major industry bodies, often representing the viewpoints of a number of stakeholders, who have the resources to keep the content of a suite of standard form contracts under constant review. Such a review process means that new editions of contract suites are published periodically, addressing statutory and case law changes, as well as addressing any perceived gaps or ambiguities in the original drafting.

The recent overhaul of the New Engineering Contract (NEC4) published in June 2017 provides a recent example. The updated suite, as described on the ICE website (Institution of Civil Engineers 2017), aims to

> support the on-going drive for further collaboration and integration of teams, greater use of modern work methods, better avoidance of disputes and more effective identification and management of risk and opportunity.

[2]Also called 'special condition'.

Guidance for Using Suitable Forms of Contract

The choice of contract flows from the procurement route chosen for the project. The choice of procurement route depends on the client's required balance of time/cost/quality and an analysis of how that can be achieved.

This must be considered in the context of the client's other requirements, not the least being his required level of involvement in the design and construction process and the extent to which he may change the specification during construction.

While these forms offer many advantages, the key is to use them correctly.

The forms can also vary in complexity, with many publishers providing forms for small projects, projects of limited scope and even forms with specific environmental or sustainability goals. Publishers also draft forms in a series that are meant to work together.

Mixing and matching incompatible forms, failing to review the incorporated additional documents, or simply using the wrong form for the parties or project involved can be problematic.

Standard forms of contract can save time and money, but should be reviewed periodically by a legal professional and read carefully in the context of each project to make sure the right form is being used, with the needed modifications, and that all of the agreements within a given project are consistent with each other.

There are various advantages to using standard forms of contract, with a wide array of standard forms available to suit the type of client, works and procurement routes involved in virtually all construction and engineering projects.

A standard form will always need to be tailored to include a project's specific details and requirements. Amending a standard form will provide an opportunity to achieve objectives while still benefiting from the generally accepted 'standard' wording in the rest of the contract since standard forms use tried and tested wording. This reduces negotiation time, cost and subsequent disputes, as most practitioners are familiar with them (Godwin 2012).

Other criteria on which a client may decide between different forms of construction contract include:

- Commercial certainty.
- Management effectiveness.
- Legal consequences.

The main factors that are considered for the selection of the form of contract are:

- Speed in terms of design and construction.
- Cost certainty.
- Dealing with complexity.
- Client's involvement.
- Capacity for variations.
- Clarity of remedies.
- Separation of design and management.

It is often the case that the only additions and changes needed are the formation of the 'Contract Particulars', also called 'Particular Conditions' and 'Contract Data', i.e. the portion of the standard form allocated to details such as party names, completion dates, liquidated damages rates, insurance details and special clauses related to specific criteria of the works (Close 2017).

An example of Contract Data is shown in Table 4.1. This is usually added as an appendix in the contract superseding the general conditions.

Advantages

Standard forms of construction contracts are ubiquitous in the industry since they have great advantages; for example:

1. They set out the rights and obligations of the parties.
2. They ensure that the risk of the project is allocated to the party that can best manage that risk.
3. They are cost-effective.
4. The terms have been vetted over years of drafting and redrafting.
5. Most clients, design professionals and contractors understand the agreement terms, or at least how the terms operate in the day-to-day realities of a building project.

In addition to the above, there are some obvious benefits in using standard forms of contract. These include:

- There is no need to produce (and incur the legal costs of producing) ad hoc contracts for every project.
- There is a degree of certainty regarding the interpretation of the clauses of the contract.
- Clauses may have been tested in the courts.
- The parties know (with reasonable certainty) the consequences of various possible courses of action.

Disadvantages

The key disadvantage to using a standard form of contracts is the necessity of using the right form for the right job, making sure that it is modified as required by the law of the country where the project is located.

The more complex the project, or the more expensive, the more important it is to review and customise the forms to fit the needs of the job.

Most standard forms are subject to significant rewrites every few years, but even in the interim periods, law changes may affect clauses' interpretations, making them unenforceable, and can subject a party to unforeseen liability.

Table 4.1 Contract data

Item	Sub-Clause	Data
Documents forming the contract listed in order of priority	1.1.1	• Form of agreement • Appendix to the agreement *(Appendix 1)* • Particular conditions attached as *Appendix 2* to the agreement • General conditions attached as *Appendix 3* to the agreement • Specifications attached as *Appendix 4* to the agreement • Drawings attached as *Appendix 5* to the agreement • Bills of quantities attached as *Appendix 6*[a]
Commencement date	1.1.7	14 days from signing the agreement[b]
Time for completion	1.1.9	12 months from the commencement date
Law of the contract	1.4	The laws of the country
Language	1.5	English
Authorized person (the employer's representative)	3.1	
Name and address of employer's representative (the project manager/consultant)	3.2	
Name and address of the contractor's representative	4.2	
Performance security	4.4	5% of the contract sum[c] stated in the agreement
Requirements for the contractor's design	5.1	N/A
Programme: time for submission and form of programme	7.2	Within 7 days of the commencement date and using the Gantt chart
Amount payable due to failure to complete the works within the time for completion	7.4	USD xx per day, but not to exceed 10% of the contract price
Period of notifying defects	9.1 and 11.5	365 days calculated from the date stated in the notice under sub-clause 8.2
Payment terms	11.2 and 11.3	Monthly progress invoices based on a percentage of completion as per the approved inspection requests
Retention	11.3	5–10% of each interim payment certificate

(continued)

Table 4.1 (continued)

Item	Sub-Clause	Data
Advance payment amount	11.3	(xx)% of the contract sum stated in the agreement against the bank guarantee
Currency of payment	11.7	USD
Insurances:	14.1	*Amount cover*:
Contractors' all risk (CAR) insurance Third-party injury to persons and damage to property Workmen's compensation		The sum stated in the agreement To be agreed As per country law

This table is generic and is included here only to show an example of how contract data is used
[a]The appendices in this table to be used as an example only. The numbering and their nominations can change according to the contract manager in charge of drafting the contract
[b]The dates and number of days are only indicative and can be changed as per agreement of the parties
[c]Performance security is typically 5% of the contract amount

Some common mistakes in using standard forms of contracts include:

1. General contractors failing to use subcontractor agreements that are compatible with the language of the standard forms and that pass through the required obligations, such as insurance limits, adding additional liability and requiring consolidated arbitration or using no written subcontractor agreements at all. This puts the general contractor in breach of the client agreement before construction even starts.
2. Making major modifications to a standard form, such as the AIA A101[3] and the A201,[4] without making any needed revisions to the incorporated documents.
3. Simply using the wrong form, such as hiring a structural engineer using an architect's contract form without changing the standard scope of work, thus making the engineer, at least technically, responsible for the architect's management of construction.

Given the complexity and specialisation of many new buildings and designs, contractors must often rely on specialist trades or products, sometimes in novel and untested ways. Therefore, even though standard language forms are often very thorough, it does not mean that they cannot be improved upon given the nature of a project.

Clients and developers may be willing to take a risk on new building design or materials in order to be at the cutting edge. In these situations, the standard allocation of risk in standard forms of contract may not reflect who should be bearing the liability

[3]AIA A101–2017 is a standard form of agreement between owner and contractor for use where the basis of payment is a stipulated sum or fixed price. A101 adopts by reference, and is designed for use with, AIA Document A201–2017, General Conditions of the Contract for Construction.

[4]AIA A201–2017 is adopted by reference in owner/architect, owner/contractor, and contractor/subcontractor agreements in the Conventional (A201) family of documents.

should something go wrong, and protective clauses should be added to the standard forms in terms of liability, design and the roles of the professionals[5] who are included.

That said, any modification to the standard language of a form should be closely scrutinised. The benefits of industry-standard language used in many form contracts are that the meaning of the language is well understood, not only by the parties to the contracts but also by the subcontractors, materials suppliers, manufacturers and even the insurance companies.

On this last point, it is very important to review changes to the standard language in these contracts from the point of insurability. Changes to the standard of care, especially regarding the performance of design professionals, could render certain contract clauses uninsurable, which rarely if ever benefits any party to the contract.

Key Tips on Amending a Standard Form: Special Conditions

Properly drawn construction contracts are vital to ensure that major projects can overcome issues or conflicts that arise during the life of the project. Contracts should provide the framework and details for the rights and responsibilities of all parties. Comprehensive and well-designed contracts can minimise the chances of any dispute requiring litigation or arbitration to be resolved.

Amending standard forms of contract is done by creating Special Conditions, also called Particular Conditions, which form a set of conditions that supplement or add to the standard forms of contract as published—called General Conditions.

Special conditions supersede the general conditions and are considered more important in terms of reliance upon them in case of a conflict between the conditions as set in the Special Conditions and General Conditions.

When amending standard forms:

- Only make amendments that are strictly necessary to comply with the parties' requirements.
- If the guidance notes to the standard form recommend a particular way of making amendments, follow the recommended practice wherever commercially possible.
- If the client is making amendments, remember these may have the effect of increasing the contract price.
- Note that amendments to one clause may have a consequential effect on other clauses including the clause numbering.
- Note also that amendments to the conditions of contract may necessitate amendments to ancillary documents such as agreements for lease, financing documents, bonds and guarantees.

---

[5]Engineer, architect, consultants and client.

- Always attach to the contract copies of the agreed forms of ancillary documents.
- Avoid letters of intent wherever possible and, if it is not possible to do so for commercial reasons, make sure the letters can be enforced subject to a financial cap.
- If amending a subcontract, remember to delete inappropriate main contract terms which may otherwise be incorporated by reference.
- Make sure the subcontract contains appropriate limitations of liability, normally by reference to the subcontract price.
- Take legal advice on complex amendments or when you are unsure of the effect of an amendment.
- Make sure that the appendices or contract and other essential contract documents are properly completed; otherwise, the contract may be unworkable or ineffective.
- If a contract has to be varied after it has been signed, make sure the variation is made by authorised representatives of the parties in accordance with the requirements, if any, specified in the contract.

Standard Forms of Contracts in Use

While there are a number of different organisations that offer standard forms of contract, they all serve the same basic function—that is, to provide an off-the-shelf form with many useful clauses that can be modified for use on specific projects.

The following standard forms of contract will be reviewed in this chapter:

1. *USA* The American Institute of Architects (AIA), ConsensusDocs and Design-Build Institute of America (DBIA) documents;
2. *UK* Institution of Civil Engineers Conditions of Contracts (ICE), New Engineering Contracts (NEC) and The Joint Contracts Tribunal (JCT) documents;
3. *International projects* These projects mostly use forms available from the International Federation of Consulting Engineers (FIDIC).

In addition, many government agencies either draft their own standardised contract forms or adapt published forms for the agencies' specific use (Godwin 2012).

The American Institute of Architects (AIA)

AIA contract documents comprise nearly 200 forms and contracts that define the relationships and terms involved in design and construction projects.

Prepared and used by the AIA with the consensus of clients, contractors, attorneys, architects, engineers, and others, the documents have been finely tuned during their 120-year history. As a result, these comprehensive contracts and forms are now widely recognised as the industry standard.

AIA Contract Documents are divided into six alphanumeric series by document use or purpose:

- *A-Series* Client/Contractor Agreements
- *B-Series* Client/Architect Agreements
- *C-Series* Other Agreements
- *D-Series* Miscellaneous Documents
- *E-Series* Exhibits
- *G-Series* Contract Administration and Project Management Forms.

The main attractions of using AIA Documents are:

- Contract terms are easy to understand.
- The contract thoroughly addresses unexpected issues that may arise in a project.
- There is a great deal of case law available as a reference to address provisions that may not be understood by parties.
- AIA contracts are easy to get and are inexpensive.

A101, A102 and A201 are considered the most highly used sets of contracts of AIA documents:

- *A101*-The standard agreement between a client and contractor for a project with a fixed-amount or lump sum payment;
- *A102*-An agreement for projects contracted for the cost of work plus a fee;
- *A201*-The general terms and conditions for the A101 and A102/A103.

The latest updates of A101, A102 and A201 were released in 2017. The revisions to A101 and A102 contained some elements of change, such as:

- Contract time provisions were added to both, to determine the official commencement of work and when the work will be substantially completed.
- Both include a new provision for payment of a termination fee if the contract is cancelled for the client's convenience, but removed the right of the contractor to recover payment for anticipated overhead and profit of uncompleted work.
- The contracts specify standards for the transmission of digital information, including building models.
- The A101 changed the requirements for progress payments that require the client to pay a portion of the contract sum allocable to completed work and materials and equipment stored at the site, while the A102 retains language referring to percentages of completion and a schedule of values.
- The A102 has added provisions related to cost controls and client approval of additional costs.
- The A102 defines what costs of the contractor are reimbursable as costs of the work, and removes some types of employee compensation—bonuses and incentives—from the list of reimbursable costs.

Some of the most important changes, however, were made to the A201:

- Insurance requirements have been moved from Article 11 of the A201 and incorporated into Exhibit A of the construction contracts. This is meant to increase flexibility and make it easier for parties to customise insurance requirements.
- Requirements for the client to provide evidence of financial arrangements to satisfy obligations under the contract have been strengthened and clarified.
- The contractor is now obliged to propose alternate means and methods if there are safety objections to client-required means and methods, rather than wait on instructions from the architect.
- There are additional requirements for scheduling, including additional milestone dates and more details about the completion of each portion of work.
- Language has been added requiring that the architect, in serving as role as the Initial Decision Maker, 'shall not show partiality to the client or the contractor'.
- The A201 adds a new severability clause that allows the bulk of the contract to remain in force even if a court determines that one or more provisions are unenforceable.

ConsensusDocs

In September 2007, a broad-based consortium of construction industry trade and other organisations jointly released a new family of model contract documents—as competitors to the AIA documents—called ConsensusDOCS.[6]

The ConsensusDOCS were not designed to favour one party over another, but rather to take a balanced approach to the rights and obligations, as well as to risk allocations, imposed on participants shall not show partiality in the construction process.

The ConsensusDOCS emphasise communication and collaboration among all project participants from the time of contract negotiation up to the completion of the project.

The stated goal of the ConsensusDOCS participants and endorsers was to reduce transaction costs and the time required for the negotiation of contracts and to reduce the frequency and severity of disputes during the construction process.

The documents are structured in numerical series by project delivery method, as follows:

- *200-series* for traditional contracting or design-bid-build.
- *300-series* for integrated project delivery.

[6]Approximately 20 construction associations, including the Associated General Contractors of America, Associated Builders and Contractors, The American Subcontractors Association, The Construction Users Roundtable, The National Roofing Contractors Association, The Mechanical Contractors Association of America, and the Plumbing-Heating-Cooling Contractors Association participated in developing the ConsensusDOCS, and endorsed them.

- *400-series for* design-build.
- *500-series* for construction management (at-risk).
- *700-series* for subcontracting.
- *800-series* for programme management and construction management (agency).

ConsensusDocs 200—Agreement and General Conditions Between Client and Contractor (Lump Sum)—is the most used. Some general characteristics of the ConsensusDocs 200 are:

1. It integrates the general terms and conditions with the contractual Agreement.
2. It emphasises the primacy of the client–contractor relationship and focuses on clear communication pathways as well as developing and maintaining positive relationships.
3. It clarifies that the client is responsible for design and design coordination; while the contractor is responsible for design elements only if specifically noted. In that situation, the client should supply all performance and design criteria.
4. It defines overhead (section 2.4.12) in a more detailed and clear manner to assist in finalising change orders and the associated costs (section 8.3.1.3) that minimise disputes during the course of the project.
5. It provides a clear and extensive definition of the cost of the work, even though this is a lump sum agreement to facilitate potential change orders without disputes.
6. It establishes criteria to set dates of substantial completion and final completion.

The main differences between ConsensusDocs and AIA documents can be summarised as follows:

- ConsensusDocs requires the client to provide information about project financing anytime upon the contractor's request.
- ConsensusDocs emphasises direct communication between the client and contractor.
- ConsensusDocs provides mutual indemnification obligations between the client and contractor.
- ConsensusDocs requires the client and contractor to negotiate a change order and a resolution for an adjustment in time or money when there is a change directive.
- ConsensusDocs provide a multi-step process by which the contractor and client first attempt to negotiate and resolve a dispute. Under the AIA, the Initial Decision Maker (IDM) makes the initial decision when there is a dispute. Under both forms, mediation is then required before the parties utilise either binding arbitration or litigation.

Design-Build Institute of America (DBIA)

When design-build construction appeared, professional associations such as the American Institute of Architects, the Association of General Contractors of America, and the Engineer's Joint Contract Documents Committee acknowledged that new contracts were needed to address this new mode of construction. They issued their own design-build contracts, but those drafted by the DBIA were more widely adopted because they were easy to follow, well-organised, and appeared to make a reasonable allocation of risk between the parties.

DBIA contract documents and forms help guide parties through the entire design-build process. From preliminary agreements to final payment, DBIA's contracts and forms are fair-basis documents that can be edited to suit a project's needs.

The DBIA contracts are fairly flexible, can be used in a wide range of design-build projects, and are appropriate for different industries. DBIA contracts assume that the project client will engage the design-contractor to carry out preliminary design and studies that end in a proposal to put together the construction documents and build the project. They consist mainly of three documents:

1. The preliminary agreement.
2. Standard agreement between client and design-contractor (a lump sum payment or cost-plus-fee with an option for guaranteed maximum price).
3. Standard form of general conditions.

DBIA forms consist of the following documents: 520, 525, 530, and 535.

The main terms that differentiate DAIB from AIA and ConsensusDocs are:

- *Ownership of documents* The design-contractor owns the product of their design efforts, while the project client receives a limited licence allowing them to use the documents under specific circumstances. If the client terminates the business relationship and continues the project with another design professional, they must pay the design-contractor a premium. The client also cannot use the design as a prototype for other projects.
- *Change* The client may make reasonable changes to the project. Change orders are used when both sides agree on a time and price adjustment to perform the work. The client may also issue a work change directive when all parties agree that a change will be made but have not yet agreed on time or pricing adjustments.
- *Hazardous conditions* If pre-existing hazardous conditions are discovered at the construction site, the contractor must stop work and notify the client, who is required to remedy the situation and compensate the contractor for any time or pricing impacts due to the delay.
- *Standard of care and warranties* The contractor must meet the standard of care necessary to achieve the agreed-upon performance standards. They agree to correct any work for a one-year period from the date that construction is substantially completed.
- *Right to stop work* The contractor may stop work and terminate the agreement if the client takes certain actions that amount to default. This includes failure to

make payments when they are due. The contractor must provide the client with a seven-day notice before stopping work.

- *Dispute resolution* In DBIA contracts, dispute resolution is a five-step process. After submitting a written notice within 21 days after the events giving rise to the dispute, the design-contractor and the client must attempt to resolve the issue themselves. If this does not work, they must meet with senior representatives of both sides within 30 days. The matter is then submitted to nonbinding mediation and, if the disagreement persists, to binding arbitration.

ICE Conditions of Contract[7]

This is a family of standard conditions of contract for civil engineering works published in the UK by the Institution of Civil Engineers (ICE).[8]

The ICE Conditions of Contract, which have been in use for over 50 years, were designed to standardise the duties of contractors, clients and engineers and to distribute the risks inherent in civil engineering to those best able to manage them. Civil engineering work is fundamentally different from building work as it involves fewer trades but can be much bigger in scale. As well, there is greater uncertainty in civil engineering works, particularly in groundworks.

The first edition of the ICE Conditions of Contract was published by the institution in 1945, in association with the Federation of Civil Engineering Contractors (now the Civil Engineering Contractors Association (CECA)). Subsequent editions followed in association with CECA and the Association for Consultancy and Engineering (ACE).[9]

In recent years the 5th (1973), the 6th (1991) and the 7th (1999) editions were published.

The most recent (7th) edition, published in 1999, was drafted by a consortium of clients, consultants, and contractors to provide a simple and standardised contract specifically tailored for civil engineering projects. It has been endorsed by the sponsoring bodies, namely the Institution of Civil Engineers, the Association for Consultancy and Engineering and the Civil Engineering Contractors Association (Haidar and Barnes 2017).

The 7th edition is based on the traditional pattern of engineer-designed, contractor-built works with valuation by measurement. This edition of the ICE Conditions introduced a number of changes, including:

---

[7]These are now renamed as the Infrastructure Conditions of Contract.

[8]ICE is an independent engineering institution and represents approximately 80,000 civil engineers worldwide. Principal membership is in the United Kingdom, but it has memberships in China, Hong Kong, Russia, India and roughly 140 other countries.

[9]In 2011, ICE withdrew from the ICE Conditions of Contract and its clientship was transferred to the Association for Consultancy and Engineering (ACE) and the Civil Engineering Contractors Association (CECA), and the contracts were rebranded as the Infrastructure Conditions of Contract.

- Incorporating some of the concepts of the Latham Report.[10]
- Amending certain provisions of the 6th edition that had attracted criticism.
- Rectifying conspicuous omissions from the text of earlier editions of the contract.
- Correcting small errors and faults from the previous edition.
- Modernising certain provisions and terms.

The ICE Conditions of Contracts continues to be the dominant form of contract for civil engineering, despite the growing importance of the New Engineering Contract.

The traditional ICE form (Client Designed Works) has strong differentiating factors from other forms:

- When a site investigation takes place for building work, boreholes and trial pits usually give a good indication of the extent of groundwater, rock and the like. This in turn means that structural work needs to be varied. The bill of quantities therefore has two main functions: firstly, as a tendering document; and secondly, as a basis for valuation.
- Substantial discretion is vested in the engineer regarding many aspects of the works.
- Price, certainly at tender stage, can be given only to the rates at which particular activities will be carried out.
- Full details of site inspection are required by the contractor prior to starting works.
- The contractor shall be deemed to have based his tender on his own inspection and the information made available by the employer or obtained by the contractor, and to have satisfied himself as to the correctness and sufficiency of the rates and prices stated by him in the bill of quantities.
- The programme should be submitted by the contractor within 21 days of the award of the contract. There are procedures for the engineer to accept or reject the programme for reasons. In that case, the contractor must amend and resubmit the programme.
- The contractor shall start the works as soon as is reasonably practicable after the works commencement date. The contractor should proceed with due expedition and without delay.
- The contractor should give notice within 28 days if the matters listed entitle him to an extension of time.
- If the engineer considers that the contractor's rate of progress is too slow, he may issue a notice to the contractor. The contractor shall then take steps to expedite the progress of the works. This is not acceleration.
- If the contractor informs the engineer that a rate for work not the subject of the variation is rendered inapplicable by the variation, the Engineer shall 'fix' the rate.
- It is a measure and value contract, so procedures for remeasurement need to be included.

[10]In 1994 the Latham Report *Constructing the Team* was published. The report was commissioned by the UK government to investigate the perceived problems with the construction industry.

New Engineering Contracts (NEC)

NEC is a generic name for a family of contracts published for the Institution of Civil Engineers by Thomas Telford Limited in the UK.

NEC stands for New Engineering Contract and it is by this acronym that the contracts are generally known. The main contract and subcontract were first published as consultative editions in 1991. First formal editions were issued in 1993, second editions in 1995, and third editions in 2005 (with updates in 2011).

It was always intended that there would be a family of New Engineering Contracts, and in the period of time between 1991 and 2005, other forms of the contracts were produced by Thomas Telford so that by 2005 the NEC3 family of contracts comprised an Engineering and Construction Contract, an Engineering and Construction Subcontract, a Professional Services Contract, a Short Contract, a Short Subcontract, an Adjudicator's Contract, a Term Services Contract, and a Framework Contract (Haidar and Barnes 2017).

The NEC3 Engineering and Construction Contract is the core document from which the options A–F are extracted. It contains all core clauses and secondary option clauses, together with the schedules of cost components and forms for contract data:

- *Option A* for priced contract with activity schedule;
- *Option B for* priced contract with bill of quantities;
- *Option C* for target contract with activity schedule;
- *Option D for* target contract with bill of quantities;
- *Option E for* cost reimbursable contract;
- *Option F for* management contract.

After 12 years of publishing NEC3, NEC introduced its fourth edition, NEC4, in 2017. NEC4 contracts reflect procurement and project management developments and emerging best practice, with improvements in flexibility, clarity, and the ease of administration.

NEC4 is an evolution on the successful NEC3. NEC4 contracts keep on using plain English and the present tense to facilitate the application of its contracts across the world. Two brand new contracts have been added to the NEC4 suite:

- NEC4 Design-Build and Operate Contract (DBO)—allows clients to procure a more integrated whole-life delivery solution; and
- NEC4 Alliance Contract (ALC)—it will support clients wishing to fully integrate a multi-party delivery team for large complex projects.

NEC contracts are used for nearly all projects procured by national and local government bodies and agencies. This is because of their unique features. For example, they:

- stimulate good management of the relationship between the two parties to contract;
- can be used in wide types of construction works and locations; and
- are clear, simple and written in plain English.

The core drafting principles of NEC4 are as follows:

- a good stimulus to good management;
- support the changing requirements of users;
- achieve transparency and certainty;
- enhance project management tools and mechanisms;
- achieve a balanced risk allocation;
- achieve clarity, transparency and certainty; and
- reflect current international best practice.

If implemented well, by parties committed to its use and to understanding how it is intended to work, the NEC offers considerable potential advantages over the more traditional families of forms.

It is not just the novelty of the NEC forms that should cause new users to be wary. The assumptions that underlie the form are in many respects fundamentally different from those in other standard forms. These are not so much questions of legal wording (though there can be much debate over that); rather, the NEC form requires of project participants a completely different philosophy towards management of the project, and therefore towards resourcing the project teams.

Perhaps the most significant feature in an NEC contract is a stronger emphasis on programming and contractual provision for regular project management meetings, and to that extent a form of early warning.

As the history of NEC has been more client-oriented, it is arguable that many of its provisions enjoy a higher level of clarity of definition than those traditionally found in other forms of contracts.

Joint Contracts Tribunal (JCT)

The Joint Contracts Tribunal (JCT) was established in 1931 and has produced, for 90 years, standard forms of contracts, guidance notes and other standard documentation for use in the construction industry, principally in the UK. In 1998, the JCT became incorporated as a company limited by guarantee, and commenced operation as such in May 1998.

Currently, JCT forms require the agreement of eight constituent bodies before they are issued by the JCT. The said constituent bodies are intended to be reasonably representative of the interests across the UK construction industry—namely, the clients, the consultants, the contractors and the subcontractors—and the JCT forms are naturally a reflection of these competing interests.

The JCT Standard Building Contract is designed for large or complex construction projects where detailed contract provisions are needed and procured via the traditional or conventional method.

A benefit of the JCT Forms of Contract is that they are tried and tested. There is, therefore, a great deal of certainty regarding the meaning of various clauses and as to how the courts will interpret those clauses (particularly as many of those clauses have at some time or another been considered by the courts).

The traditionally perceived problem with the JCT Forms of Contract was that they were not flexible enough to cope with modern requirements. Many of these perceived problems, however, were dealt with by the publication in 2005 (updated in 2011 and 2016) of an entirely new suite of contracts and subcontracts.

Some of the more commonly used JCT contracts are:

- Standard Building Contract.
- Intermediate Building Contract.
- Minor Works Building Contract.
- Design and Build Contract.
- Constructing Excellence Contract.

The JCT has progressed through various editions in providing for greater proactivity on management of change. There is a discernible trend in modern construction procurement towards a prior assessment of the impact, in cost and in time, of a proposed change in the scope of the works (variation), before its implementation. Some of the main characteristics of a JCT contract are:

1. Under JCT Contracts, the contract administrator typically administers the contract on behalf of the employer and has a duty to act impartially between parties. The primary emphasis of the contract administrator is on payment administration and guiding the parties through the procedures and processes under the contract.
2. Contract ground risk is with the contractor.
3. Its emphasis is on the submission of a master programme, and only if an extension of time is requested does it allow for a revised programme.
4. It is mostly based on lump sum contract with or without quantities
5. The JCT contract contains provisional sums.
6. In relation to payment, the JCT contract payment section is clear, is all in one section (clause 4), and is easy to follow.
7. It contains comprehensive detail in relation to insurance, in clause 6 and Schedule 3.
8. It has a clear interaction between the employer's requirements and contractor's proposals, and there is extensive drafting in the JCT
9. On first impression, there may be a superficial and misleading similarity in appearance between a JCT contract administrated with bills of quantities, and an ICE contract. Both involve interim measurement as the monthly mechanism for ascertaining interim amounts payable to the contractor. The prevalence of interim measurement (despite recent trends to milestone and S-curve payment systems) continues to dominate British construction industry practice.
10. Under a JCT contract, the interim measurement is merely an administrative, provisional way of allowing payments on account of what will ultimately be a lump sum, and therefore in that sense, fixed price.

Federation Internationale Des Ingenieurs-Conseils (FIDIC) Contracts

The Fédération Internationale Des Ingénieurs-Conseils (FIDIC) was founded in Belgium in 1913. Since then, it has become the foremost representative body for the world's consulting engineers, with member associations in some 100 countries. Although the Contracts Committee became one of its earliest constituent parts soon after FIDIC's foundation, it was not until 1957 that the first FIDIC standard form contract was produced.

FIDIC is recognised as the leading body for the development of model standard forms of contract for use in the international construction industry. The standard forms are generally accepted by clients and contractors as providing a balanced allocation of risks and providing fair procedures for the administration of contracts.

FIDIC also produces standard forms of contract for civil engineering construction that are used throughout the world. FIDIC contracts are often referred to, with very good reason, as the international standard form of contract.

In the mid-1990s, two significant landmarks occurred in the history of the development of the FIDIC contracts. These were the introduction of a turnkey contract, the Orange Book, and the setting up of a task group to produce a major revision of the Red and Yellow Books.

These events led to the launch in 1999 of the so-called 'Rainbow Suite' from the colours of the covers of the respective Books: Red, Yellow and Silver.

The new FIDIC forms have adopted much of current thinking in the administration of contracts and have attempted to balance the risk and responsibility between the client and the contractor.

In December 2017, 18 years after FIDIC released its First Edition Rainbow Suite in 1999, FIDIC published second editions of the Red, Yellow and Silver Books among others as updates to the first editions.

FIDIC is usually divided into two parts:

- *Part I* contains the general terms of the contract, such issues as rights and obligations of each party, procedure for payment, variation, certification and dispute resolution.
- *Part II* of the contract is the conditions of particular application and is to be used to introduce project-specific clauses, such as language of the contract, choice of law, and the name of the person or firm appointed to act as Engineer or Client's representative for the project, among other terms.

In most FIDIC forms there is a default hierarchy for the documents forming the contract. The order of priority as stated below shows a standard example of how the priority of documents is usually set, and in the event of inconsistency the first on the list takes precedence (Robinson 2011):

- The contract agreement.
- The letter of acceptance (this is the formal acceptance of the contractor's tender and marks the formation of the contract).

- The letter of tender.
- Part II—the conditions of particular application.
- Part I—general conditions of contrac.
- The specification and drawings.
- The client's requirements.
- The schedules.
- Further documents (if any), listed in the contract agreement or in the letter of acceptance.

The parties are allowed to rearrange the priority of documents or stipulate that no priority or order of hierarchy will apply to the contract. This can be done in Part II of the contract.

The components of the present suite of FIDIC contracts are often best known by their colours, and the latest contracts are:

- The Red Book, which is the Conditions of Contract for Construction.
- The Yellow Book, which is the Conditions of Contract for Plant and Design-Build.
- The Silver Book, which is the Conditions of Contract for EPC/Turnkey Projects.
- The Green Book, which is the Short Form of Contract.
- The Blue Book, which is the Form of Contract for Dredging and Reclamation Works (the 'Dredgers Contract').
- The White Book, which is a form of agreement for engagement of Consultants.

An Overview of the Leading FIDIC Contracts: Red, Yellow and Silver Books

The new suite of FIDIC contracts, published in 2017, comprises new editions (described by FIDIC as second editions) of the Red, Yellow and Silver Books. Conceptually, the new versions are similar to their 1999 predecessors: the Red Book is FIDIC's 'traditional procurement' client design contract, the Yellow Book has the dual function of design-and-build/contractor design and mechanical/electrical plant procurement, and the Silver Book is FIDIC's EPC/Turnkey Contract (Robinson 2011).

The Conditions of Contract for Construction for Building and Engineering Works Designed by the Client: The Red Book

The Red Book is not only the oldest of the FIDIC contracts, celebrating its 60th anniversary in 2017; it is also the most widely used for general construction projects of many kinds. The single most important characteristic of the Red Book is contained in its full title—it is a client design contract.

The design, prepared by the Client's staff or by consultants acting on its behalf, is provided to the contractor in the form of specifications and drawings (and any schedules). The payment mechanism is traditional measurement and valuation. It is, however, open to the parties to vary this position.

The main characteristics of the Red Book are:

- FIDIC advises of 'the possibility of replacing Clause 12 by appropriate Particular Conditions for a lump sum contract or a cost-plus contract'.
- The role of the engineer is traditional: 'who shall carry out duties assigned to him by the Contract'.
- It may seem paradoxical that the engineer's duties, including such important functions as the making of determinations, should be usually much more fully set out in the contract, to which the engineer is not a party, than in the contract for professional services (or employment contract) which governs the relationship between client and engineer.
- This is regarded as normal in common law jurisdictions, but the 'dual role' of the engineer as agent of the client, and as a decision maker acting fairly between the parties, is sometimes regarded with misgivings in civil law jurisdictions.
- In other respects, the FIDIC Red Book follows in the line of its predecessors as an engineering contract in the common law style.

The Conditions of Contract for Plant and Design-Build for Electrical and Mechanical Plant and for Building and Engineering Works Designed by the Contractor: The Yellow Book

The Yellow Book is FIDIC's second-oldest contract, and almost certainly its second most widely used, for electrical and mechanical plant and also for design and build works more generally (Baker et al. 2013).

As with the Red Book, the contract's most important feature is contained within the full title—it is a design and build contract. The design is prepared by the contractor in accordance with the client's requirements, which specify 'the purpose, scope and/or design and/or other technical criteria for the works.'

The Yellow Book has long been one of FIDIC's most important contracts. The content of its earlier editions influenced the Silver Book to some extent and also the Gold Book, both of which are based on the contractor design concept. The main characteristics of the Yellow Book are:

- In the Yellow Book, the contractor accepts a fitness-for-purpose obligation for the works, including the design.
- The payment mechanism for the Yellow Book is lump sum fixed price, with provision for progress payments on the basis of engineer certification.
- Like the Red Book, the Yellow Book contract is administered by the engineer (see above).

- A major difference is evident from the Clause 20 dispute resolution provisions. Whereas the Red Book provides for a 'standing' Dispute Adjudication Board (DAB) to be appointed '*by the date stated in the Appendix to Tender*,' under the Yellow Book, the parties '*jointly appoint a DAB by the date 28 days after a Party gives notice to the other Party of its intention to refer a dispute to a DAB*'.
- Its risk allocation (see Chapter 5, 'Allocation of Risk in Construction Contracts') bears more resemblance to the Red Book in terms of the perception of traditional balance between Client and Contractor.

The Conditions of Contract for EPC/Turnkey Projects: The Silver Book

The FIDIC Silver Book is the third of the major Rainbow Suite contracts. It was the most controversial upon its launch and to some extent remains so. This is largely attributable to perceptions of its risk allocation.

FIDIC's first turnkey contract was the Orange Book but the Silver Book was seen as a greater departure from the traditional FIDIC contracts which preceded the 1999 Rainbow Suite.

Essentially, the Silver Book is a lump sum Engineering–Procurement–Construction (EPC) turnkey contract. As with the Yellow Book, the design is prepared by the contractor in accordance with the client's requirements. The contractor assumes full responsibility for the engineering, procurement and construction of the works and undertakes a fitness-for-purpose obligation for the Works, including the design.

This form of contracting is typically used on complex engineering facilities, such as process or power plants, where a high degree of certainty as to cost, time and performance is required, often because of 'bankability' issues in funding the project. The concept is that the Client obtains a fully functioning facility, capable of operating immediately to guaranteed standards of performance, i.e. ready at the 'turn of a key'.

While the Silver Book, in keeping with the EPC turnkey concept, allocates much more risk to the Contractor than in more 'balanced' contracts, such as the Red and Yellow Books, it should not be regarded as controversial. Contractors routinely do price for the varying degrees of risk allocation in contracts.

The introductory note of the Silver Book states that the aim of the Book is to give certainty as to the time and costs of the works (Baker et al. 2013).

The main features of the Silver Book are that:

- The Contractor takes full responsibility for the design of the project.
- The project is on a lump sum basis.
- The possibilities for the Contractor to request and obtain adjustment of the price are limited.

Other main characteristics of the Silver Book are as follows:

- Particular attention should be given to that which is stated in the contract and in the technical specifications to be implied by the contractor's obligations, as specified under Clause 4 and, in particular, in Sub-Clause 4.1 [Contractor's General Obligations]: '*When completed, the Works shall be fit for purposes for which the Works are intended as defined in the Contract*'.
- With specific limited exception, the employer shall not be responsible for any error, inaccuracy or omission contained in the Employer's Requirements as included in the Contract.
- The administration of the contract remains in the hands of the employer unless the latter appoints, under Clause 3.1 [*The Employer's Representative*] the employer's representative (which appointment is therefore optional). The employer's representative does not, however, play an independent role. The representative is expressly the *longa manus* of the employer.
  - There are three main types of variations:
  - those purely instructed by the Employer;
  - those proposed by the Contractor upon request of the Employer;
  - those proposed by the Contractor.
- The Silver Book aims at providing the Employer certainty in terms of costs and time for completion; however, it provides certain cases where the contractor is entitled to obtain costs, profit and/or extension of time.

It is fair to say that the provisions of the Silver Book are slightly more favourable to the Employer. The use of the Silver Book (but this is true also for the other FIDIC forms) requires, however, a competent contract administration both from the employer and from the contractor to avoid any pitfall and to ensure that it reaches fully the aim the parties have in mind (Hewitt 2014).

Other FIDIC Contracts: Specialised Works

FIDIC—Subcontract Contracts

The latest FIDIC Conditions of Subcontract for Construction (First Edition, 2011) for building and engineering works designed by the Employer has been prepared for use with the FIDIC Red Book fourth edition. The General Conditions are prepared to operate 'back to back' with the 1999 Red Book in terms of rights and obligations.

The Subcontract carries a fitness-for-purpose obligation, to the extent that the Subcontractor is responsible for design. Payment is by measurement in accordance with the Main Contract (i.e. Red Book) provisions.

In the case of the Force Majeure provisions, this is literally done with the words 'The provisions of Main Contract Clause 19 (Force Majeure) shall apply to the Subcontract.'

The form is also noticeable for a series of flow charts. These represent typical sequences of the principal events, of payment events and of Subcontractor claims and disputes under alternative versions of the dispute resolution provisions.

Obviously, there is no engineer appointed, and contract administration is in the hands of the contractor. Subcontract disputes are resolved by reference to a subcontract DAB, followed by time for amicable settlement in the event of reference to arbitration, which provides for International Chamber of Commerce (ICC) Rules with one arbitrator rather than three as the default position.

FIDIC: Consultancy Agreements[11]

FIDIC consultancy agreements differ from the Rainbow Suite Contracts and the other standard in that they are not construction contracts and do not involve the contractor (Hewitt 2014).

Most significant of these by far is the White Book. First published in 1990, the White Book is probably the most significant agreement for professional services in construction globally. The contract is between the 'purchaser' of the services: 'Client' rather than 'Employer', and the 'supplier' of the services: 'Consultant' rather than 'Contractor'.

FIDIC's intention was to produce an agreement for general use for the purposes of pre-investment and feasibility studies, detail design and administration of construction and project management, both for client-led design teams and for contractor-led design teams on design and build commissions.

A feature of the 2006 (and previous editions) of the White Book was the very limited provision concerning the rights and obligations of the Parties. The basic Consultant's duty of care was stated to be '*no other responsibility than to exercise reasonable skill, care and diligence in the performance of his obligations under the Agreement*'. This was consistent with the English law concept of the basic duty of the supplier of a service.

An attempt has been made in the 2017 fifth edition to make this more flexible, by providing that the 'Consultant shall perform the Services with a view to satisfying any function and purpose that may be described in Appendix I [Scope of Services]'.

However, there has been a significant strengthening of the reasonable skill, care and diligence standard itself to 'that to be expected from a consultant experienced in the provision of such services for projects of similar size, nature and complexity'.

The actual substance of the scope of services is very limited. Some more guidance notes have been inserted in the 2017 fifth edition; but the Appendices are still

[11] This section quotes from Baker and Lavers (2017).

largely blank pages, to be filled in by the parties. Consequently, what are in effect further Particular Conditions will usually be much greater in volume than the General Conditions.

The fifth edition has also added new provisions relating to client-instructed variations to the services. A mechanism exists for instructing variations but '*any such Variation shall not substantially change the extent or nature of the Services*'.

The dispute resolution provisions of the White Book have always differed sharply from those found in the construction contracts. Under the 2017 fifth edition, disputes are first subject to an amicable dispute resolution stage, which is now stated to be mandatory, followed by adjudication with a further provision for amicable settlement before the reference to arbitration, either subject to agreement by the Parties or ICC Rules.

In addition to the fifth edition of the White Book, in 2017, FIDIC published second editions of its Model Joint Venture (Consortium) Agreement and Sub- Consultancy Agreement, both developed for use with the White Book.

Short Form of Contract (First Edition, 1999) Green Book

These Conditions of Contract are recommended for engineering and building work of relatively small capital value. However, depending on the type of work and the circumstances, the conditions may be suitable for contracts of considerably greater value.

The Green Book is considered most likely to be suitable for fairly simple or repetitive work or work of short duration without the need for specialist subcontracts. This form may also be suitable for contracts that include, or wholly comprise, contractor-designed civil engineering, building, mechanical and/or electrical works.

References

Baker E, Lavers A (2017) Introduction to the FIDIC suite of contracts. In: Brekoulakis S, Brynmor D (eds) Guide to construction arbitration. Law Business Research Ltd, London

Baker E, Mellors B, Chalmers S, Lavers A (2013) FIDIC contracts: law and practice. Taylor and Francis, Abingdon

Close JA (2017) Comparative guide to standard form construction and engineering contracts. Law Brief Publishing, Minehead

Godwin W (2012) International construction contracts. Wiley-Blackwell, Hoboken, NJ

Haidar A, Barnes P (2017) Delay and disruption claims in construction, 3rd edn. ICE Publishing, London

Hewitt A (2014) The FIDIC contracts obligations of the parties. Wiley-Blackwell, Hoboken, NJ

Institution of Civil Engineers (2017) Changes in NEC4 works contracts. https://www.ice.org.uk. Accessed 16 Jan 2021
Klee L (2017) International construction contract law, 2nd edn. Wiley-Blackwell, Hoboken, NJ
Robinson MD (2011) A contractor's guide to the FIDIC conditions of contract. Wiley-Blackwell, Hoboken, NJ

Chapter 5
Contract Drafting and Main Conditions

Abstract This chapter considers the skills that a contract manager must possess in order to draft a contract and the essential points he/she must consider in drafting a well-structured contract between an employer and the contractor. As well, this section will guide contract managers when reviewing a contract already prepared by others. Major clauses in a contract are listed and each clause is described to give the reader the basic knowledge needed in drafting or reviewing a construction contract and what he/she must consider as a priority. This chapter examines the principles and the main essential clauses that should be taken into consideration when drafting a construction contract. In essence, this chapter complements Chapter 4, which reviews the standard forms of contract, as each standard form includes many of the headings listed in this chapter and approaches it differently, in many instances. A brief description is included for each clause heading and emphasis is placed on some of the main clauses that are recurrent in construction contacts. Any professional in this field can research each topic in depth if he so requires.

Keywords Dispute resolution · Drafting contracts · Rights and remedies · Termination · Time for completion · Variations and change orders

Drafting

The actual works on all construction projects exhibit recurrent distinctive characteristics such as:

- The nature and scope of the works.
- Specification and design details.
- Interactivity between employer, contractor and supplier.
- Provisions for changes.
- Complexity of sequencing of activities and dependencies on other activities/supplies.
- Site characteristics.

A. D. Haidar, *Handbook of Contract Management in Construction*,
https://doi.org/10.1007/978-3-030-72265-4_5

- Outside factors such as fluctuations in raw materials prices, lack of resources due to demand and Acts of God[1].
- Product's life.
- Responsibility and liability of parties towards defects (Scott 2016).

Drafting a contract in construction takes the form of art, skill, knowledge and expertise and must include the scope of works as well as responsibilities of the parties and their limitations (Scott 2016). Hence, in order to draft a well-structured contract, the contract manager must have full knowledge of:

1. The expectation of the parties.
2. The project's scope.
3. Commercial details including variation orders and payments.
4. Complete knowledge of time and schedules.
5. General conditions such as default, termination, suspension, insurance, warranties.
6. Dispute resolution process.

Bad drafting may cause unnecessary disputes. Some steps that must be considered by the contract manager in drafting a contract, and to minimise disputes, are (Kenneth 2018):

1. Obtain full and clear instructions and information as to what the contract aims to achieve.
2. Avoid drafting a clause without being able to foresee all possibilities intended to be covered.
3. Keep instructions and information made during negotiations and refer to them when needed to make sure that the draft is made according to the intention of the parties.
4. Become knowledgeable in the subject matter and investigate with the parties all relevant issues that might arise during the performance of the works.
5. Make a structured plan of the documents to be inserted in the contract.
6. Use published forms and forms from previous documents wherever possible.
7. Avoid using unclear clauses.
8. Make as many revisions of the draft as necessary.
9. Take one definition/clause at a time and check that it fully and consistently deals with the various sections of the document.
10. Special alterations or additions made to the standard forms must be explicit and precise.
11. Use simple and clear words and easily understandable sentences.

[1] *Force majeure.*

Considerations in Drafting a Contract

The following key considerations must be taken by an experienced contract manager when drafting a contract:

- The terms of the contract are comprehensive and fair to both contracting parties.
- The legitimate interests of both contracting parties are appropriately considered and balanced.
- The legitimate interests of each party include the right to enjoy the benefits of the contractual relationship.
- Best practice principles of fair and balanced risks/rewards allocation between the employer and the contractor are put into effect.
- No party shall take undue advantage of its bargaining power.
- Provisions for the contractor/subcontractor to be paid adequately and promptly are included in the contract to maintain its cash flow while the employer obtains the best value for money.
- To the greatest extent possible, cooperation and trust between the contracting parties are promoted, and contra proferentum is avoided.
- The contract provisions can be put into practice.
- A proper mechanism for disputes is inserted so disputes are avoided as much as possible, and resolved efficiently.

Main Conditions in Contracts

The purpose of this section is to give direction and examples for the main terms in a contract, and to provide some of their definitions. All clauses definitions and descriptions in the remaining of the section are only indicative and must be used as a guide to the reader.

Each clause mentioned, herein, shall require modifications and rewriting to allow for the project scope; the statutory laws of the country; and the parties' requirements such as exceptions, commercial terms and so on. As well, these conditions must be complementary to the standard forms of contract to be used for each project.

This section has the following headings as they are the most essential in any contract:

1. Adjudication
2. Basic Definitions
3. Employer Responsibilities
4. Engineer
5. Contractor Responsibilities
6. Design, Specifications and Documents
7. Time for Completion
8. Payments
9. Variations and Change Orders

10. Indemnifications
11. Claims
12. Dispute Resolution
13. Warranty
14. Works
15. Termination
16. Suspension
17. Others (Injury or Damage to Person or Property, Governing Law, Notices, Rights and Remedies, Mutual Responsibility, Invalidity, Entire Agreement).

It is to be noted that many of the provisions described herein are also described in other chapters of this book.

Basic Definitions

Contract

The contract represents the entire and integrated agreement between the parties and supersedes prior negotiations, representations or agreements, either written or oral.

Contract Documents

The contract documents consist of the agreement between the employer and the contractor, conditions of the contract, drawings, specifications, addenda issued prior to execution of the contract, instructions to bidders, bill of quantities, schedule of rates and other documents listed in the agreement.[2]

Contractor

The person or entity identified as such in the agreement, and referred to throughout the contract documents. Unless otherwise stated, the term 'Contractor' in a contract means the general contractor or the general contractor's authorised representative.

[2]In the event of conflicts among the contract documents, documents with the order they are listed takes precedence.

Drawings

The drawings include the design, location and dimensions of the works. They also include plans, elevations, sections and diagrams.

Engineer

The engineer, as identified in the contract. The engineer can be the contract manager or administrator in certain cases.

Employer

The person or entity identified as such in the agreement, and referred to throughout the contract documents. The term 'Employer' in a contract means the employer or the employer's authorised representative.

Project

The project means the works performed under the contract documents as a whole or as a part, and which may include construction works carried out by the employer or by separate contractors.

Specifications

The specifications are that portion of the contract documents consisting of the written requirements for materials, equipment, construction systems, standards and workmanship for the works and performance of related services.

Work(s)

The term 'work or works' means the construction and services required by the contract documents, whether completed or partially completed and includes all labour, materials, equipment and services provided or to be provided by the contractor to fulfil the contractor's obligations. The works may constitute the whole or a part of the project.

Employer's Responsibilities

The employer's main responsibility is to secure and pay for the works executed, and the necessary approvals, easements and charges required for construction.

Without cost to the contractor, the employer shall furnish to the contractor, when required, a survey describing the physical characteristics, legal limits and utility locations for the project site.

All other information or services under the employer's control shall be furnished by the employer within a reasonable time after a written request from the contractor to avoid delay in the orderly progress of the works.

The employer's duties and responsibilities, that are mandatory in a typical project, include:

1. Financing for the project.
2. Site surveys.
3. Securing and paying for easements.
4. Providing the plans and specifications.
5. Furnishing the materials in a timely fashion.
6. Disclosing superior knowledge.
7. Acting on clarifications and changes.
8. Interpreting the documents.
9. Cooperating with the contractor.
10. Selecting all professionals on the project.
11. Interacting with various governmental bodies.

Engineer[3]

Engineer's Administration of the Contract

The engineer shall provide administration of the contract as described in the contract documents, and shall be the employer's representative during construction until the completion of works. The engineer shall have authority to act on behalf of the employer only to the extent provided in the contract documents unless otherwise modified in writing.

The engineer shall supervise construction means, methods, techniques, sequences, procedures and safety precautions and he shall deliver required programmes in connection with the works.

[3]Further reading in Chapter 7.

The engineer shall not be responsible for the contractor's failure to carry out the works in accordance with the contract documents; neither shall he be responsible for acts or omissions of the contractor, subcontractors, or their agents or employees, or of any other entities performing portions of the works.

Except as otherwise provided in the contract documents or when direct communications have been specially authorised, the employer and the contractor shall endeavour to communicate through the engineer. Communications by and with subcontractors and material suppliers shall be through the contractor.

Based on the engineer's inspections, observations and evaluations of the contractor's payment certificates, the engineer shall review and certify the amounts due to the contractor, and shall issue certificates for payment in such amounts.

The engineer shall have the authority to reject any works which do not conform with the contract documents. Whenever the engineer considers it necessary or advisable for implementation of the intent of the contract documents, the engineer shall have authority to require additional inspection or testing of the works, whether or not such works are fabricated, installed or completed.

Neither the authority of the engineer, however, nor a decision made in good faith either to exercise or not to exercise such authority shall give rise to duty or responsibility of the engineer towards the contractor, subcontractors, material and equipment suppliers, their agents or employees, or other entities performing portions of the works.

Review by the Engineer

The engineer shall review and approve or take other appropriate action upon the contractor's submittals such as shop drawings, materials submittals and samples, but only for the limited purpose of checking for conformance with the information given and the design concept expressed in the contract documents.

The engineer action shall cause no delay to the works or to the activities of the employer, contractor or separate contractors, while allowing sufficient time in his professional judgement to permit adequate review.[4]

The engineer's review of the contractor's submittals shall not relieve the contractor of his obligations.

[4]Review of submittals by the engineer is not conducted for the purpose of determining the accuracy and completeness of other details such as dimensions and quantities, or for substantiating instructions for installation or performance of equipment or systems, all of which remain the responsibility of the contractor as required by the contract documents.

Engineer's General Duties

The engineer shall prepare change orders and construction change directives, and may authorise minor changes in the work without obtaining the employer's approval.

The engineer shall conduct inspections to determine the date or dates of substantial completion and the date of final completion. He/she shall receive and forward to the employer, for the employer's review and records, written warranties and related documents required by the contract and assembled by the contractor, and shall issue a final certificate for payment upon compliance with the requirements of the contract documents.

The engineer shall interpret and decide matters concerning performance and requirements under the contract documents on the written request of either the employer or the contractor. The engineer response to such requests shall be made with reasonable promptness and within any time limits agreed upon.

Interpretations and decisions of the engineer shall be consistent with the contract documents and shall be in writing or in the form of drawings. When making such interpretations and decisions, the engineer shall endeavour to secure faithful performance by both employer and contractor and shall not show partiality to either of them.

If the contractor fails to correct works that are not in accordance with the requirements of the contract documents, or persistently fails to carry out any works in accordance with the contract documents, the engineer may order the contractor to stop the works, or any portion thereof, until the cause for such order has been dealt with properly.[5]

Contractor's Responsibilities

Contractor's Responsibilities and Construction Procedures

The contractor shall carefully study and compare the contract documents and all information furnished by the employer, and shall at once report to the engineer any errors, inconsistencies or omissions discovered.[6]

The contractor shall be solely responsible for and have control over construction means, methods, techniques, sequences and procedures, and shall be responsible for coordinating all portions of the works under the contract unless the contract documents give other specific instructions concerning these matters.

[5] The right of the engineer to stop the work shall not, however, give rise to a duty on the part of the engineer to exercise the right for the benefit of the contractor or any entity.

[6] If the contractor, while performing a construction activity, is either aware or should have been aware that it involves an error, inconsistency or omission in the contract documents, and does not give notice of such to the designer, the contractor shall assume full responsibility for such performance and shall bear the full costs for correction.

The contractor shall be responsible to the employer for acts and omissions by the contractor's employees, subcontractors and their agents and employees and other entities performing portions of the works under a contract with the contractor.

The contractor shall not be relieved of his obligations to perform the works in accordance with the contract documents either by activities or duties of the engineer done by the engineer's administration of the contract, or by tests, inspections or approvals required or performed by entities other than the contractor.

The contractor shall be responsible for inspection of portions of the works already performed under the contract to determine that such portions are in proper condition to receive subsequent works.

Labour and Materials

Unless otherwise provided in the contract documents, the contractor shall provide and pay for labour, materials, equipment, tools, construction equipment and machinery, water, heat, utilities, transportation and other facilities and services necessary for proper execution and completion of the works, whether temporary or permanent and whether or not incorporated or to be incorporated in the work.

The contractor shall enforce strict discipline and good order among the contractor's employees carrying out the contract, and shall not permit the employment of unfit persons or persons not skilled in tasks assigned to them.

Design, Specifications and Documents

Drawings and Specifications

Drawings, specifications and other documents prepared by the engineer are instruments of the engineer's services regarding the works to be executed by the contractor. Neither the contractor nor any subcontractor, or any other party affiliated to the contractor shall own or claim copyright of the drawings, specifications and other documents prepared by the designer.

The employer is usually the only party that has the right to retain the copyright of the drawings, specifications and other documents prepared by the designer.

The drawings, specifications and other documents prepared by the engineer, and copies thereof furnished to the contractor, are for use solely with respect to the project. They are not to be used by the contractor or by any subcontractor or material or equipment supplier on other projects without the specific written consent of the employer and engineer.

Documents and Samples

The contract documents shall be signed by the employer and the contractor as provided in the agreement. The contractor shall maintain at the site, for the employer's use, at least one record copy of the drawings, specifications, addenda, change orders and other modifications in good order and marked currently to record changes and selections made during construction. As well, the contractor shall maintain shop drawings, product data, samples and similar required submittals.

These shall be available to the engineer and shall be delivered to the engineer for submittal to the employer upon completion of the work.

Shop Drawings, Product Data and Samples

Shop drawings are drawings, diagrams, schedules and other data specially prepared for the works by the contractor, subcontractors, manufacturer, supplier or distributor to illustrate a specific activity of the works.

Product data are illustrations, standard schedules, performance charts, instructions, brochures, diagrams and other information furnished by the contractor to illustrate materials or equipment for some part of the works.

Samples are physical examples, which illustrate materials, equipment or workmanship and establish standards by which the work shall be judged.

The purpose of preparing shop drawings, product data, samples and similar submittals is to demonstrate, for those portions of the work for which submittals are required, the method that the contractor proposes in order to conform to the information given and to the design concept expressed in the contract documents; this purpose shall be reviewed by the engineer.[7]

The contractor shall review, approve and submit to the engineer shop drawings, product data, samples and similar submittals required by the contract documents, with the following conditions:

- Submittals are done with reasonable promptness and in such sequence as to cause no delay to the activities of the employer or of separate contractors.
- The contractor shall perform no portion of the works requiring these submittals until each respective submittal has been reviewed and approved by the engineer.
- The contractor represents that he/she has determined and verified materials, field measurements and field construction criteria related thereto and has checked and coordinated the information contained within such submittals with the requirements of the works and the contract documents.
- The contractor shall not be relieved of responsibility for deviations from the requirements of the contract documents by the engineer's review and approval of these submittals unless the contractor has specifically informed the engineer

[7] These are subject to the limitations of the contract.

in writing of such deviation at the time of submittal and the engineer has given written approval to the specific deviation.

Time For Completion

Construction Programmes[8]

Unless otherwise provided, contract or project time is the period of time allotted in the contract documents for substantial completion of the work.

The date of commencement of the works is the date established in the agreement. The date shall not be postponed by the failure to act by the contractor or other entities for whom the contractor is responsible.

The date of substantial completion is the date certified by the engineer in accordance with the contract provisions.

Immediately after being awarded the contract, the contractor shall prepare and submit for the engineer's review and comment a construction schedule for the works.[9]

The schedule shall in principle:

1. not exceed time limits provided in the contract documents;
2. be revised at appropriate intervals as required by the conditions of the works;
3. be related to the entire project to the extent required by the contract documents; and
4. provide for expeditious and practicable execution of the work.

The contractor shall prepare and keep current, for the engineer's approval, a schedule of submittals that is coordinated with the contractor's construction schedule and allows the engineer reasonable time to review submittals.

Progress and Completion

By executing the agreement, the contractor confirms that the contract time is a reasonable period for performing the works. The contractor shall be liable for and shall pay the employer such sums as shall be set forth in the agreement between the employer and the contractor as liquidated damages[10] for delays until the works are substantially complete.

The contractor shall not, except by agreement or instruction by the employer in writing, prematurely commence operations on the site or elsewhere prior to the effective date of insurance required by the contract for construction to be furnished by the contractor.

---

[8] Also called schedules.

[9] Called baseline schedule or baseline programme.

[10] Usually not to exceed 10% of contract sum.

Unless the date of commencement is established by a notice to proceed given by the engineer, the contractor shall notify the employer and the engineer in writing not less than five days[11] before commencing the works.

The contractor shall proceed expeditiously with an adequate number of direct and indirect personnel, and shall achieve substantial completion within the contract time.

Delays and Extensions of Time[12]

If the contractor is delayed at any time during the works by an act or neglect of the employer or engineer, or by changes ordered in the works, or other causes beyond the contractor's control, then the contract time shall be extended for such reasonable time as the engineer may determine.

If the works are interrupted or hindered by the employer or the engineer, the contractor shall be entitled to an extension of time in an amount equal to such interruption or hindrance.

Payments

Contract Sum

The contract sum (lump sum) as stated in the agreement is the total amount payable by the employer to the contractor for performance of the works under the contract documents.

At least fifteen (15) days[13] before the date established for each progress payment, the contractor shall submit to the engineer an itemised application for payment[14] for works completed in accordance with the schedule of values.[15] Such application shall be supported by all data substantiating the contractor's right to payment as the employer or engineer may require: copies of requisitions from subcontractors and material suppliers and reflecting retention sum if provided for elsewhere in the contract documents.

Unless otherwise provided in the contract documents, payments shall be made on account of materials and equipment delivered and suitably stored at the site for subsequent incorporation in the works. If approved in advance by the employer, payment may similarly be made for materials and equipment suitably stored off

[11]Or similar as inserted in the agreement.

[12]Further readings in Chapters 6 and 7.

[13]This can vary according to the parties' agreement.

[14]Also called 'payment certificate'.

[15]Part of the bill of quantities (BOQ).

the site at a location agreed upon in writing. Payment for materials and equipment stored on or off the site shall be conditioned upon compliance by the contractor with procedures satisfactory to the employer to establish the employer's title to such materials and equipment or otherwise protect the employer's interest, and shall include applicable insurance, storage and transportation to the site for materials and equipment stored off the site.

The contractor warrants that title to all the works covered by an application for payment shall pass to the employer no later than the time of payment.

The contractor further warrants that all the works for which certificates for payment have been issued and payments received from the employer shall, to the best of the contractor's knowledge, be free and clear of liens, claims, security interests, or encumbrances.

Provided an application for payment is received by the engineer not later than the 'agreed ' day of a month, the employer shall make payment to the contractor not later than the agreed time for payment as stipulated in the contract.

Advance Payment

The employer shall make an advance payment, in most contracts, as an interest free loan for the works, when the contractor submits a guarantee in an approved form.[16]

The engineer shall issue an interim payment certificate for the advance payment, and after that, the employer shall receive a guarantee for the amounts and currencies equal to the advance payment. The guarantee shall be issued by a bank approved by the employer, and shall be in the form attached to the contract or in another form approved by the employer.

The contractor shall ensure that the guarantee is valid and enforceable until the advance payment has been repaid, but its amount may be progressively reduced by the amount repaid by the contractor as indicated in the payment certificates. If the terms of the guarantee specify its expiry date, and the advance payment has not been repaid by the expiry date, the contractor shall extend the validity of the guarantee until the advance payment has been repaid.

The advance payment shall be repaid through percentage deductions from the payment certificates.

[16]Further readings in Chapter 7.

Retention

To ensure proper performance of the contract, the employer shall retain five per cent (5%)[17] of the amount of each approved application for payment until the project works are complete, provided that the contractor continues to perform satisfactorily and any nonconforming work identified in writing prior to that date has been corrected by the contractor and accepted by the employer.

Certificates for Payment

The engineer shall, within seven days (or as stipulated in the contract) after receipt of the contractor's application for payment, either issue to the employer a certificate for payment, with a copy to the contractor, for the amount that the engineer determines is properly due, or notify the contractor and employer in writing of the engineer's reasons for withholding certification in whole or in part.

The engineer's certification for payment shall constitute a representation to the employer, based on the engineer's inspections at the site and on the data comprising the contractor's application for payment, that the works have progressed to the point indicated and that the inspections of the construction, repairs, or installations have been conducted with the degree of care and professional skill and judgement ordinarily exercised by a member of his profession.

The engineer's certification for payment shall be signed and sealed by the engineer and presented to the employer. The issuance of a certificate for payment shall further constitute a representation, by the engineer, that the contractor is entitled to payment for the amount certified.

Progress Payments

After the engineer has issued a certificate for payment, the employer shall make the payment in the manner and within the time provided in the contract documents, and shall so notify the engineer.

The contractor shall promptly pay each subcontractor, upon receipt of payment from the employer, out of the amount paid to the contractor on account of said subcontractor's portion of the work, the amount to which the subcontractor is entitled, reflecting percentages actually retained from payments to the contractor on account of the subcontractor's portion of the works.[18]

[17] 5% is standard and could be up to 10% in specific projects.

[18] The contractor shall, by appropriate agreement with each subcontractor, require each subcontractor to make payments to sub-subcontractors in similar manner.

Schedule of Rates

Before the first invoice for payment, and if otherwise required by the contract documents, the contractor shall submit to the employer a schedule of values allocated to the various portions of the works, prepared in such form and supported by such data as to substantiate said schedule's accuracy as the employer may require.

This schedule, unless objected to by the employer, shall be used only as a basis for the contractor's invoice for payment and to asses variations.

Failure of Payment

If the engineer does not authorise payment or notify the contractor in writing of the employer's reason for withholding approval, through no fault of the contractor, within forty-five (45) days (or as noted in the contract) after receipt of the contractor's invoice for payment, or if the employer does not pay the contractor within fifty (50) days (or as noted in the contract) after the date any amount was authorised by the employer's procurement representative, the contractor may, upon ten (10) additional days' written notice (or as noted in the contract) to the employer representative, stop the works until payment of the amount owed has been received.

Variations and ChangeOrders

Change Orders Protocol

Changes and variations to the works may be instructed during the execution of the contract, and without invalidating the contract, by change order, construction change directive,[19] or an order for a minor change in the works, subject to the limitations stated in the contract and elsewhere in the contract documents (Calamari et al. 2018).

- the following variations to the works may be permissible: A change order shall be based upon agreement among the employer, the contractor, and engineer;
- a construction change directive requires agreement by the employer and the engineer and may or may not be agreed to by the contractor; and
- an order for a minor change in the work may be issued by the engineer alone.

Changes in the works shall be performed under applicable provisions of the contract documents, and the contractor shall proceed promptly unless otherwise provided in the change order, construction change directive, or order for a minor change in the work.

[19] Also called 'site instruction'.

If unit prices are stated in the contract documents or subsequently agreed upon, and if quantities originally contemplated are changed in a proposed change order or construction change directive, such unit prices shall be applied to the quantities of the works proposed.

Overhead and profit shall be in the proximity of 10% (or as agreed) of the value of labour and materials for works performed by the contractor, or 5% (or as agreed) if the works are performed by his subcontractors.[20]

A change order is then issued by the engineer and signed by the employer, the contractor and the engineer, stating their agreement upon all of the following:

- A change in the works.
- The amount of the adjustment in the contract sum, if any.
- The extent of the adjustment in the contract time, if any.

Construction Change Directive

The employer may, without invalidating the contract, order changes to the works within the general scope of the contract consisting of additions, deletions or other revisions. This will have the effect of the contract sum and contract time being adjusted accordingly through a construction change directive procedure.

A construction change directive[21] is a written order prepared by the engineer and signed by the employer and the engineer, directing a change to the works and stating a proposed basis for adjustment, if any, in the contract sum or the contract time, or both.

If the construction change directive provides for an adjustment to the contract sum, the adjustment shall be based on one of the following methods:

- Mutual acceptance of a lump sum properly itemised and supported by sufficient substantiation to allow a proper evaluation.
- Unit prices stated in the contract documents or subsequently agreed upon; Cost to be determined in a manner agreed upon by the parties and with a mutually acceptable fixed or percentage fee.

Upon receipt of a construction change directive, the contractor shall promptly proceed with the change to the works involved and advise the engineer of the contractor's agreement or disagreement with the method, if any, provided in the construction change directive for determining the proposed adjustment in the contract sum or the contract time.

A construction change directive signed by the contractor indicates the agreement of the contractor therewith, including adjustment to the contract sum and the

[20] If the works are performed by a subcontractor, the main contractor's overhead and profit shall not exceed 5% (or as agreed).

[21] Also called 'work order'.

contract time or the method for determining them. Such agreement shall be effective immediately, and shall be recorded as a change order.

If the contractor does not respond promptly or disagrees with the method of adjustment to the contract sum, the method and the adjustment shall be determined by the engineer on the basis of reasonable expenditures and savings for the works related to the change, including, in case of an increase in the contract sum, a reasonable allowance for overhead and profit (Calamari et al. 2018).

Following the engineer's issuing of a change order directive, the contractor shall keep and present, in such form as the engineer may prescribe, an itemised accounting with appropriate supporting data which include the following:

- Costs of labour.
- Costs of materials, supplies and equipment, including cost of transportation.
- Hired costs of machinery and equipment.
- Costs of premiums for all bonds and insurance, permit fees and taxes.
- Additional costs of supervision and field office personnel directly attributable to the change.
- Additional time.

When the employer and the contractor agree with the determination made by the engineer concerning the adjustments in the contract sum and the contract time, or otherwise reach an agreement upon the adjustments, such an agreement shall be effective immediately and shall be recorded by preparation and execution of an appropriate change order.

Minor Change Order

Change orders that relate to minor works, and are of relatively low monetary value, can be instructed by the engineer without the employer's approval.

The contract must allow such a responsibility to the engineer so that he can proceed with issuing a minor change order without hindering the works. This can take the form of site instruction by the engineer to gain time and proceed quickly with certain tasks.

The same procedure of change order directive must be followed to assess the quantities, value and time needed for these minor works.

Indemnification

A general indemnity clause is usually included in the agreement that states:

> To the fullest extent permitted by law, the contractor shall indemnify and hold harmless the employer, engineer, engineer's consultants, and agents and employees or any of them from

> and against claims, damages, economic losses and expenses of any kind (including but not limited to fees and charges of engineers, attorneys, and other professionals and costs related to court action or arbitration), arising out of or resulting from performance of the work under the agreement, provided such claim, damage, loss or expense is attributable to bodily injury, sickness, disease or death, or to injury to or destruction of tangible property (other than the work itself) including loss of use resulting therefrom, caused in whole or in part by negligent acts or omissions by the contractor, a subcontractor, anyone directly or indirectly employed by them or anyone for whose acts they may be liable unless caused in whole or part by the negligence of the employer. Such obligation shall not be construed to negate, abridge, or reduce other rights or obligations of indemnity which would otherwise exist as to a party or person described in the contract.

The obligations of the contractor under the indemnity clause shall not extend to the liability of the engineer, the engineer's consultants, and agents and employees of any of them arising out of:

- The preparation or approval of drawings, reports, surveys, change orders, designs or specifications;
- The giving of, or the failure to give, directions or instructions by the engineer, the engineer's consultants, and agents and employees of any of them, provided such giving or failure to give is the primary cause of the injury or damage.

Claims[22]

A claim is a demand or assertion by one of the parties seeking, as a matter of right, adjustment or interpretation of contract terms, payment of money, extension of time or other relief with respect to the terms of the contract.

The term 'claim' can also include other disputes and matters in question between the employer and the contractor arising out of or relating to the contract. Claims must be made pursuant to the dispute resolution procedure as set in the contract. The responsibility to substantiate claims shall rest with the party making the claim (Nancy 2016).

If the contractor wishes to make a claim for an increase in the contract sum, a written notice shall be given before he proceeds to execute the work.

If the contractor believes the additional cost is involved for reasons including but not limited to:

- A written interpretation from the engineer.
- A written order for a change in the works issued by the engineer.
- Force majeure.
- Delays due to employer causes.
- Design deficiency or incomplete design.
- Delayed payments by the employer; Contractual breaches by the employer and/or the engineer then a claim shall be submitted with full substantiation of the relevant

[22] See Chapter 7 for further details.

facts. The contractor's claim shall include an estimate of the costs and of probable effect of the delays to the progress of the works.

Dispute Resolution

Dispute Resolution Procedure

To prevent all disputes and litigation, it is usually agreed by the parties that any claim, question, difficulty or dispute arising from the agreement in place or the construction process shall be first submitted to the engineer to address the issue (Nancy 2016).

Upon review of the claim, the engineer shall take one or more of the following preliminary actions within fourteen (14) days (or similar as noted in the contract) of receipt of a claim:

- Request additional supporting data from the claimant.
- Submit a schedule to the parties indicating when the engineer expects to act.
- Reject the claim in whole or in part stating the reasons for rejection.
- Recommend approval of the. claim by the other party; or
- Suggest a compromise.

Although he is not obligated to do so, the engineer may also give notification of the nature and amount of the claim to the surety or the bank (if any) providing the bonds and guarantees.

If a claim has not been resolved within a specified period in the contract, the party making the claim shall, within fourteen (14) days[23] after the engineer's preliminary response, take one or more of the following actions:

- Submit additional supporting data requested by the engineer.
- Modify the initial claim and resubmit it to the engineer.
- Notify the engineer that the initial claim stands and submit the claim to mediation or arbitration.[24]

[23] Or as noted in the contract.

[24] Dispute resolution clauses in standard forms of contract often stipulate that in the event a dispute arises, parties have to undertake certain steps such as negotiation and mediation to resolve the dispute amicably before commencing arbitration. Before discussing how domestic courts approach these multi-tiered clauses, it is useful to consider the advantages and the disadvantages of such clauses.

Mediation

The mediator shall address any properly submitted claim, question, difficulty or dispute arising from the agreement or the construction process that has not been satisfactorily resolved by the engineer. Such requests shall be made to the employer in writing after the engineer's preliminary response.

The mediator shall notify the contractor in writing of the decision within, usually, twenty eight (28) calendar days from the date of the submission of the claim to the mediator, unless the mediator requires additional time to gather information or allow the parties to provide additional information.

The mediator's orders, decisions and decrees shall not be binding. Mediation shall be a pre-condition to initiating litigation or arbitration concerning the dispute. The mediation session shall be private. Prior to the commencement of mediation, if requested by either party or the mediator, the parties and the mediator shall execute a written confidentiality agreement in accordance with the provisions of applicable law.

If the disputed issue cannot be resolved in mediation or either party disagrees with the results of the mediation, the parties may seek a resolution by arbitration or in courts as the contract dictates.

Some advantages of a mediation approach before arbitration are:

- It saves the parties the expense of an arbitration if the dispute is settled with high-level talks or negotiations.
- It acts as a 'second opportunity' for both the parties and their advisers to re-evaluate the expense of an arbitration with regard to the outcome and the net profit or goodwill of the business.
- It helps to preserve long-term relationships between employers, contractors, engineers and other professionals and does not jeopardise future business opportunities.
- It reduces the aggregate number of issues to be resolved in an actual arbitration.[25]

Some disadvantages of a multi-tiered approach:

- Where the dispute has reached a deadlock, or where only a third-party verdict on the merits of the issues will suffice, the exercise of going through the motions of a negotiation or mediation may constitute wastage of resources.
- In urgent matters where time is of the essence, the opportunity of obtaining interim measures may be lost.
- It increases the risk of the arbitration clause being rendered ineffective. In particular, where there are no well-defined stages of negotiation or mediation, this can lead to ambiguity in determining the beginning and termination of the process and hence the start of arbitration.

[25]Through negotiation or early meditation, frivolous or trivial claims can be settled or written off at the outset.

- Its interplay with any limitation act can lead to a bar in commencing arbitration, owing to the time taken in negotiation, mediation and high-level talks.

Arbitration

Arbitration is the final procedure in a dispute.[26] In conducting arbitration between the litigant parties, the arbitrator must provide the platform whereby:

- Each party must have a full opportunity to present his own case to the tribunal.
- Each party must be aware of his opponent's case, and must be given a full opportunity to test and rebut it.
- The parties must be treated alike. Each must have the same opportunity to put forward his own case, and to test that of his opponent.

The arbitration condition contained in a contract is considered a separate agreement from the other conditions of the contract. The invalidity of the contract—which includes the arbitration condition—or its revocation or termination shall not result in the invalidity of the arbitration condition that is included in the contract if this condition is true in itself (Nancy 2016).

The use of arbitration is commonly used in the construction industry because of:

- The prevalence of arbitration clauses in standard forms of contract.
- The technical content of disputes, leading to the use of arbitrators skilled in technical disciplines; The need in many disputes for the arbitrator to be empowered to open up, review and revise decisions or certificates arising from the architect's or engineer's judgement in administering the building contract.

In modern practice, almost all arbitration agreements are set out in writing, either by way of an express dispute resolution clause in the contract or by way of some other mechanism that provides a written basis for the agreement. Hence, to proceed in arbitration, the contract must disclose that the arbitration agreement:

- is made in writing (whether or not it is signed by the parties);
- is made by an exchange of communications in writing;
- is evidenced in writing (including circumstances where an agreement made otherwise than in writing is recorded by one of the parties, or by a third party, with the authority of the parties); or
- is not made in writing but is made by reference to terms that are in writing.
- The three sets of laws that may apply to any given arbitration are:
- the proper law of the contract;
- the proper law of the arbitration agreement; and
- the proper law of the conduct of the arbitration.

[26]If no court proceedings are listed in the contract as the final procedure in a dispute.

Warranty

The contractor warrants to the employer and the engineer that:

- The materials and the equipment furnished under the contract shall be of good quality and new unless otherwise required or permitted by the contract documents.
- The works shall be free from defects not inherent in the quality required or permitted; The works shall conform with the requirements of the contract documents.

If required by the engineer, the contractor shall furnish satisfactory evidence as to the type and quality of materials and equipment, and that the works not conforming to these requirements, including substitutions not properly approved and authorised, may be considered defective.

Where the manufacturer's warranty on equipment, or parts thereof, exceeds twelve (12) months, the guarantee period on such equipment, or parts thereof, shall be extended to include the full warranty of the manufacturer.

The contractor shall repair or replace such defective materials, equipment or workmanship to the full satisfaction of the employer within the stipulated guarantee period without cost to the employer.

Works

Access to Works

The contractor shall confine operations at the site to areas permitted by law, ordinances, permits and the contract documents, and shall not unreasonably encumber the site with materials or equipment.

The contractor shall provide the employer and engineer access to the works in preparation and progress, wherever these may be located.

Uncovering and Correction of Works

If a portion of the works is covered contrary to the engineer's request or to the requirements specifically expressed in the contract documents, it must, if required in writing by the engineer, be uncovered for the engineer's observation and be replaced at the contractor's sole expense without change in the contract time.

If a portion of the works has been covered which the engineer has not specifically requested to inspect prior to its being covered, the engineer may request to see the works and it shall be uncovered by the contractor.

If the works are not in accordance with the contract documents, the contractor shall pay such costs unless the condition was caused by the employer or a separate contractor in which event the employer or separate contractor shall be responsible for payment of such costs.

If the works are in accordance with the contract documents, the employer, by appropriate change order, shall be charged with the cost of uncovering and replacement.

The contractor shall promptly correct all the works rejected by the engineer or failing to conform to the requirements of the contract documents, whether these works be observed before or after substantial completion and whether or not they be fabricated, installed or completed. The contractor shall bear any and all costs of correcting such rejected works, including additional testing and inspection, and compensation for the engineer's services and expenses made necessary thereby.

If, within one year (or as stated in the contract) after the date of substantial completion of the works or designated portion thereof, or after the date for commencement of warranties or by terms of an applicable special warranty required by the contract documents, any of the works are found to be not in accordance with the requirements of the contract documents, the contractor shall correct them promptly after receipt of written notice from the employer to do so.

If the contractor fails to correct nonconforming works within a reasonable time, the employer may correct these defective works at the contractor's expense.

Tests and Inspections

Tests, inspections and approvals of portions of the work required by the contract documents shall be made at the required time. Unless otherwise provided, the contractor shall make arrangements for such tests, inspections and approvals with an independent testing laboratory or entity acceptable to the employer, and the employer shall bear the costs of tests, inspections, and approvals.

If such procedures for testing, inspection or approval reveal the failure of portions of the works to comply with the requirements established by the contract documents, the contractor shall bear all costs made necessary by such failure, including those of repeated procedures, and compensation for the engineer's services and expenses.

Required certificates of testing, inspection, or approval shall be secured by the contractor and promptly delivered to the engineer.

If the engineer is required by the contract documents to observe tests, inspections, or approvals, the engineer shall do so promptly and, where practicable, at the normal place of testing.

Tests or inspections conducted pursuant to the contract documents shall be made promptly to avoid unreasonable delay in the work.

Termination

Contractors generally have relatively limited termination rights in construction contracts. Contractors are typically entitled to terminate a contract in the following specific events:

- a failure by the employer to submit reasonable evidence of financial arrangements demonstrating the ability to pay the contract price;
- if the employer substantially fails to perform, or materially breaches, its obligations under the contract;
- if the employer engages in corrupt or fraudulent practice in relation to the contract; For a prolonged suspension of the works instructed by the employer or his representative; and
- for employer insolvency.

Provided one of the prescribed reasons arises, and the contractual procedure is followed, a contractor will be entitled to terminate the contract.

The consequences of termination will be prescribed in the contract and will generally include payment of sums due and return of performance security to the contractor, as well as handing over of plant, materials and documents paid for by the employer and removal of all other goods from the site.

As will be further discussed in Chapter 7 of this book, employers have generally more extensive termination rights. While a valid termination for default may have disastrous consequences for contractors, a wrongful termination may entitle them to payment for all the profit they would have made under the contract but for the wrongful termination and the cost of demobilisation.

The employer may, without cause, order the contractor in writing to suspend, delay, or interrupt the works in whole or in part for such period of time as the employer may determine.

However, the employer may also terminate the contract for one of the following reasons if the contractor:

- Persistently or repeatedly refuses or fails to supply enough properly skilled workers or proper materials.
- Fails to make payment to subcontractors for materials or labour in accordance with the respective agreements between the contractor and the subcontractors.
- Persistently disregards laws, ordinances, rules, regulations or orders of a public authority having jurisdiction; Otherwise is in substantial breach of a provision of the contract documents.

When any of these reasons exist, the employer, upon certification by the engineer that sufficient cause exists to justify such action, may without prejudice to any other rights or remedies of the employer and after giving the contractor and the contractor's surety (if any) seven days' (or as stated in the contract) written notice, terminate employment of the contractor and may take possession of the site and all materials, equipment, tools and construction equipment and machinery thereon owned by the contractor, and finish the works by whatever reasonable method the employer may deem expedient.

Suspension[27]

There is a very close relationship between suspension and termination and, depending on how the clause is drafted, the end result of a suspension clause may be much the same as a termination clause in that either party will have the right to terminate the contract at the end of the agreed suspension period.

In the absence of an express contractual term, it would be difficult to argue that a general right to suspend exists in law, as the courts have consistently refused to recognise such a right. It is, therefore, recommended that the parties consider having a suspension clause in their contracts. If they do have one, they should also ensure that the contract deals adequately with the immediate practical consequences of a suspension order, and with the question of how long a contract can be suspended before termination may occur. Consideration should also be given to what happens when works are to resume following suspension.

If the contractor suffers delay and/or incurs cost from complying with the engineer's instructions under the Suspension of Work clause and/or from resuming the works, the contractor shall give notice to the engineer and shall be entitled to an extension of time for any such delay, if completion is or will be delayed, and shall also be entitled to payment of any such cost, which shall be included in the contract price.

After receiving this notice, the engineer shall proceed in accordance with the Engineer Determinations clause to agree or determine these matters.

The contractor shall not be entitled to an extension of time for, or to payment of the cost incurred in, making good the consequences either of the contractor's faulty design, workmanship or materials, or of the contractor's failure to protect, store or secure the works during the suspension.

Others

Injury or Damage to Person or Property

If either party to the contract suffers injury or damage to person or property because of an act or omission of the other party, or any of the other party's employees or agents, or of others for whose acts such party is legally liable, written notice of such

[27] For further reading, see Chapter 7.

injury or damage, whether or not it be insured, shall be given to the other party within a reasonable time not exceeding 10 days (or similar) after the first observance.

The notice shall provide sufficient detail to enable the other party to investigate the matter.

Governing Law

It is preferable that the contract is governed by and in accordance with the laws of the country in which the works are being carried out, and comprises that:

- The contract documents shall be construed, enforced and regulated under and by the laws of the country where the project site works are being performed.
- The contractor consents to the jurisdiction of the courts of the country or local institution where the project site is being performed with respect to the commencement of any legal action.

Notices

It is a standard practice that a written notice shall be deemed to have been duly served by the employer and the contractor if it is delivered in person to the individual or to a member of the firm or entity of the corporation for which it was intended, or if it is delivered at, or sent by registered or certified mail to, the business address listed in the contract.

Rights and Remedies

Duties and obligations imposed by the contract documents and rights and remedies available shall be in addition to and not a limitation of duties, obligations, rights and remedies otherwise imposed or available by law.

Mutual Responsibility

A mutual responsibility clause, in essence, works between the main contractor and his subcontractors and their relationships towards the employer. The subcontractor is required to perform its obligations consistent with the main contractor's obligations to the employer, and the subcontractor is granted the same rights against the main contractor that the main contractor has against the employer.

To the extent that the terms of the main agreement apply to the subcontract works, then the contractor assumes towards the subcontractor all obligations, rights, duties, and redress that the employer under the main agreement assumes toward the contractor (Nancy 2016).

In an identical way, the subcontractor assumes towards the contractor all the same obligations, rights, duties and redress that the contractor assumes towards the employer under the main contract.

Invalidity

The invalidity or unenforceability of all or any part of any provision of the contract documents shall in no way affect the validity or enforceability of any other provision or the remainder of any such provision.

Entire Agreement

The contract documents (including such items of the contract documents as may be issued subsequent to the execution of the contract) embody the full and complete understanding of the parties and supersede any previous agreements, written or oral, between employer and contractor, and may be modified only in writing signed by the employer and the contractor.

References

Calamari J, Perillo J, Bender H, Brown C (2018) Cases and problems on contracts, American Casebook Series, 7th edn. West Academic Press, St. Paul, MN

Kenneth AP (2018) A manual of style for contract drafting, 4th edn. American Bar Association, Chicago, IL

Nancy K (2016) The fundamentals of contract law and clauses: a practical approach. Edward Elgar Publishing, Northampton, MA

Scott JB (2016) Drafting and analyzing contracts: a guide to the practical application of the principles of contract law, 4th edn. Carolina Academic Press, Durham, NC

Chapter 6
Time and Costs—Claims

Abstract This chapter is higher-level reading material, aimed at the more senior contract managers and professionals dealing with critical and more involved issues related to time and costs and claims generally. Delay and disruption claims are described to give the reader an understanding of what they are and the situations they can be applied to. These are important in contract management, as most often delay and disruption (also called loss and expense) claims are pursued by a contractor for losses incurred as a result of the prolongation and/or disruption of the works. As well, global claims are reviewed. Techniques used in order to substantiate the claims and costs quantifications are also reviewed. Other topics such as acceleration, reasonable time, time of essence and mitigation are also reviewed to give the reader a basic understanding of these topics. They are generally complex in nature and require much expertise of any professional who wants to tackle any of them. Finally, a matrix providing loss and expense to claims is added at the end of the chapter (Table 6.1) to give contract managers a guide to identifying and calculating the relevant damages to a particular claim.

Keywords Acceleration · Claims · Delay and disruption · Global claim—total cost method · Loss of productivity · Windows analysis

Delay and Disruption

Construction and engineering projects are subject to considerable risks and uncertainties. These include weather, soil conditions, availability of labour, materials and plant and sometimes the intervention of certain government bodies and local authorities. Such uncertainties frequently cause delays in the project's scheduled programme and, ultimately, the completion of the project (Haidar and Barnes 2017).

Delays and disruptions, which often result in projects finishing late and over budget, often are supplemented by enormous claims for compensation or result in the contractor facing liquidated damages.

A. D. Haidar, *Handbook of Contract Management in Construction*,
https://doi.org/10.1007/978-3-030-72265-4_6

Many techniques are applied to analyse delays and disruptions.[1] The three principal evidential aspects to proving delay and disruption are as follows:

- The requirement to prove that the event occurred and the extent of the event and its consequential effects.[2]
- The requirement to prove that the contractor has some entitlement under the contract or that the employer is otherwise liable for the delay event.[3]
- The requirement to establish that the event had an effect on the completion date of the works.[4]

A well-structured delay and disruption claim should clearly set out the following:

- Factual events that have led to a delay or a financial loss and a statement as to why the employer is responsible for these events.
- Evidence that each of these events is covered by the contract and that any stipulations for making the claim have been complied with, such as time limits for making a claim or failure of the employer to adhere with the terms of the contract such as payments or approvals (Burr 2016).
- Evidence of a strong causal link between the events complained of and the delay or financial loss.

Any notice required to be served under the contract should make particular reference to the relevant clause number under which it is being served, and the contractor should follow precisely any procedure pertaining to the form or content of the notice. As notices generally require action by other parties, it is advisable, if not imperative, to keep a log of all notices given and of the dates by which response should be made by the employer (or the engineer).

Delay and Disruption (Loss and Expense) Claims

The expression 'delay claim' is usually used to describe a monetary claim which follows on from a delay to the works as a whole.

The expression 'disruption claim' is used to describe a monetary claim in circumstances where part of the works has been disrupted, without affecting the ultimate completion date of the project; this typically equates with the delays that are not on the critical path (Haidar and Barnes 2017).

---

[1]Delays and disruptions in construction projects are frequently expensive, since there is usually a construction leverage involved which charges interest, bonds and guarantees with ongoing fees related to the down payment and performance, management staff dedicated to the project whose costs are time dependent and ongoing inflation in wage and material prices.

[2]This is usually, but potentially not in every case, a question of fact and is largely dependent upon the records of events as they occurred.

[3]This is largely a question of law.

[4]This will often require some sort of delay of analysis of the activities involved and proper records of the data that lead to this knock-on effect.

The liabilities and costs related to delays and disruptions are frequently a subject of contention. For contractors, and frequently consultants too, these manifest themselves as liquidated or actual damages, labour, material and equipment costs, extended head and site office overheads and loss of productivity. For employers, they appear as loss of profit, revenue opportunity costs[5] and consultants' fees.

Delays and disruptions are two different types of damages. Delay damages are valid only if delays to the overall project completion time are involved, while disruption damages can be caused by any change in the planned condition of work that can happen regardless of the change in the project completion time (Burr 2016).

The priced tender has limited relevance for the evaluation of delay and disruption caused by breach of contract, or any other cause which requires the evaluation of additional costs. The tender pricing may, however, be relevant as a base line for the evaluation of delay and disruption caused by variations.

During the operation of construction contracts disputes often arise (for example because of the late completion of the works, delay in handling site handover, delay in submittals approvals, etc.), hence, the right to recover loss and expense is included in many construction contracts as being the means of recovering losses arising from a specified breach (or breaches) of contract.[6]

If the matter relied on does amount to a breach of contract, however, then damages under statutory law may or may not be applicable depending on the case and the facts. In this instance if a party to a contract breached the contract under misrepresentation, duress, frustration or any other principle, the other party can claim for damages for such a breach (Gibson 2015).

A claim for loss and/or expense is the means of putting a contractor back into the position in which he would have been but for the delay or disruption. It is not a means of turning a loss into a profit.[7]

Participants in the construction and civil engineering industries often make the mistake of believing that time and money are directly linked, in that they assume that financial claim recovery can only be secured if an extension of time has been granted and, conversely, that it will certainly be due if an extension has been given. In reality, neither of these propositions is actually correct. Additional monies may be recovered even if prolongation is not present, and alternatively, a contractor may receive a comprehensive extension of time and still not be entitled to additional financial recovery.

[5]Consequential loss is not usually allowed in any construction dispute.

[6]In some ways, loss and expense can be considered as being a contractual mechanism for the recovery of what may otherwise be considered as being damages.

[7]Therefore, where a contractor's costs are materially different from the rates quoted in the contract for preliminaries, there must be a good reason for that difference otherwise the claimed costs may be disallowed in total or in part on the grounds that the costs claimed are unreasonable.

Time in Construction

Completion Dates

Programming of the complex sequences of activities and their dependencies is one of the principal skills of the successful contractor, and is critical for the success of any project. All but the simplest of projects, however, will proceed from some such a programme (Haidar 2011).

The traditional forms of contracts make little contractual provision for integrating the programming of activities into structural obligations. The contractor's time-based obligation under the traditional forms comprises only an obligation to complete the totality of the works by a particular date.

Time is a complex parameter in the matrix of a construction contract; it often leads to delays and disruptions for the concerned project. The doctrines and principles that create causations are:

- Types of delays, time of essence and the reasonableness test.
- Extension of time.
- Completion matters, concurrency, acceleration and *time at large*.
- Project programming and float. This issue has a complex nature and is becoming critical in proving claims and assisting in the calculation of damages.
- Liquidated damages, the doctrine of *quantum meruit* and their calculations.
- Mitigation and remedies.

Hence, time-scaled networks of the as-planned, as-built, and as-adjusted schedules are used to present the facts during claims negotiation, arbitration or litigation processes and are compiled using painstaking research of the site labour, materials and other records as required to produce the global claim quantification.

Extension of Time (EOT)

The period during which the contract remains valid corresponds to the period originally set for the completion of the works. Nothing short of agreement of the parties can extend the duration and validity of the contract.

When the period fixed for the completion of the contract is about to expire, the question to grant an extension of time for completion of the works should be considered by the employer/engineer. The failure to extend the time on or before the date on which the period, whether originally fixed or extended, will render the employer unable to operate the clause relating to liquidated damages.

In a case where a contract does not contain a proper mechanism for allowing the employer to grant an extension of time for delays that are caused by the employer or by anyone or anything which the employer is responsible for, liquidated damages are irrecoverable as the revised completion date cannot be set or identified under the provisions of the contract. In other words, time becomes 'at large', through the operation of the prevention principle (the concept of 'time at large' and the prevention principle are dealt with in more detail in Chapter 7 of this book).

Where there is an extension of time clause, this is regarded as being inserted for the benefit of the employer, since it operates to keep alive the liquidated damages clause in the event of delay being due to an act of the employer or the engineer.[8]

The extension must in any case be made at a reasonable time before the time limited for completion of the works has expired (unless there is some power in the contract to extend the time after completion), so that the contractor may know the time within which he has to complete and arrange his works accordingly (Society of Construction Law 2017).

Each contract must have its own clause relating to extension of time suitable to serve the interest of the employer. If the contractor desires an extension of time for completion on the grounds that he/she has been unavoidably hindered in the execution of the works or any other ground, the contractor shall apply in writing to the engineer within 30 days[9] of the date of the hindrance on account of which he desires such extension.[10] In construction contracts, compliance with such contractual notices will usually be a condition precedent to any claim for an extension of time.[11]

- In addition to the notice requirements, the contractor will be required to follow the procedures, as set out in the contract, in order to claim an extension of time. This will generally involve submitting detailed particulars and supporting contemporaneous records evidencing the entitlement to an extension of time.[12]
- The engineer shall authorise any such extension of time that may, in his opinion, be necessary or proper after careful evaluation of the circumstances and their impact on the original completion date of the project.

A multitude of events can delay a project, some of which are outside the contractor's control. Accordingly, contractors cannot always be held responsible for delays and liable for the resulting losses. Depending on the governing law and the provisions of the agreed contract, a contractor may be entitled to claim an extension of time when such events arise. The events that can give rise to an extension of time include:

---

[8]In many cases the time fixed by the contract ceases to be applicable on account of some act or default of the employer or the engineer.

[9]Or as noted in the contract.

[10]Contractors will usually be required to provide notice to the employer, or its representative, of the delaying event. The notice must commence when the contractor becomes aware, or should have become aware of the delaying event.

[11]These contractual notices are enforced strictly in accordance with one of the fundamental principles of contract law namely *pacta sunt servanda* or freedom of contract.

[12]Large and lengthy projects will generally be divided into milestones that the contractor must achieve by set milestone dates. In such circumstances, contractors are likely to also be liable for liquidated or general damages if they fail to complete the milestones by the milestone dates.

- Acts caused by the employer or one of its representatives (including other contractors hired by the employer).
- Events outside of both parties' control, such as force majeure events, material or goods shortages or delays caused by the authorities.
- Variations to the scope of works.

The most common heads of a delay claim are prolongation, disruption[13] and acceleration. When a project is delayed because of an event for which the employer is responsible, the contractor is entitled to an extension of time, as discussed above, as well as any losses arising from the delay. These losses are referred to as prolongation costs.

Prolongation costs can include site and office overheads, financing charges arising from money borrowed to fund the project, and loss of profit. To successfully claim for prolongation costs, contractors need to provide detailed particulars and evidence that each of the claimed costs was caused by employer delay.

Types of Delays

A delay is defined as the time during which some part of the construction project is completed beyond the contractual completion date or after the date when the extension of time expires. Delay may be caused not just by the employer or contractor but by any party participating in the project, such as the designer, main contractors, subcontractors, suppliers and utility companies. A delay might also result due to *force majeure*,[14] such as a severe weather event, a fire or an earthquake.

Construction delays can be categorised as two major types; namely, excusable and non-excusable.

An excusable delay is one for which the contractor is excused from meeting a contractual completion date and for which he/she will, therefore, receive a time extension. Excusable delays can be caused either by the employer or by a third party not participating directly in the contract. In general, excusable delays include unforeseen design problems, variations and change orders, site restrictions, late payments and Acts of God such as fire, strikes and wars.

A non-excusable delay involves lost time caused directly by the contractor's actions or inactions. In this case, the contractor is entitled neither to time extension nor to additional compensation from the employer. Generally, non-excusable delays include the contractor's failure to perform work within the agreed time period, poor work performance and resource availability problems. This type of delay is also called culpable delay.

Excusable delays can be further categorised into two types, namely compensable delays and non-compensable delays. A compensable delay allows the contractor both a time extension and additional damage costs. In this case, the employer should

[13] Due to the delays.

[14] Also called Acts of God.

compensate not only for the damage costs caused by the compensable delay, but also for the cost of any follow-up work necessitated by the delay. The damages include all costs incurred by the contractor due to the delay such as overhead costs, interest on payments, and all related losses such as procurement and design contingencies. In the event of a non-compensable delay, the contractor is not entitled to compensation for additional costs caused by the delay, but may be entitled to a time extension.

Delay Claims

Methodology

Assessing the impact of a delay, and establishing a direct causal link from the delay event to the effect, liability and resulting cost and/or damages is a complex exercise. In selecting the most appropriate technique to determine a delay, the relevant facts, the timetable, and the nature and number of delays and events, as well as the size of the potential dispute to ensure that the matter is dealt with proportionately and efficiently, must all be considered.

The most important and indeed difficult task in terms of determining delay is to establish the nexus of cause and effect. The mere happening of an event that, it is considered, caused a delay, confers no entitlement to an extension to the contract completion date.

The following points must be established in determining an extension of time entitlement, and the process to be followed:

- That an event occurred at all.
- That the event occurred in the manner asserted.
- That the event falls within a category providing an express entitlement to an extension of time.
- That the required notices were given regarding the cause of delay and identifying the clause in the contract that stipulates that the specified cause entitled the contractor to an extension of time.

If the engineer (or contract manager/administrator) makes his determination, he/she is then required to offer to the contractor in writing an extension of time that he/she considers to be fair and reasonable. What constitutes the basis of a fair and reasonable extension of time was considered in *John Barker Construction Limited v London Portman Hotel Limited*,[15] where HHJ Toulson QC set out the criteria that had to apply in order to calculate a fair and reasonable extension of time that comprised the fact that the architect (in that case) recognised the effects of constructive change, made a

[15](1996) 83 BLR 31.

logical analysis, in a methodical way, of the effect of relevant events on the contractor programme and calculated, rather than made an impressionistic assessment of, the time taken up by relevant events.

Delay Substantiation

Various methods of calculating a delay have been adopted, and these vary according to the terms of the contract, the factual and analytical method used in calculating the delay, and the experts involved in assessing the claim. In essence, what is being sought is the delay that is caused by the relevant events occurring at the time when they did in fact occur, with the project being in the state that it was in at that time and the contractor responding to it as he did, with an allowance then being made for any extent to which the contractor has, through breaches of contract, contributed to the resulting delay.[16]

For contract managers and experts, the underlying question is whether or not the tribunal will be convinced, on a balance of probabilities,[17] that a relevant event caused a delay of a certain duration.

Models may be sufficient to demonstrate it but, equally, they may not. Sometimes a theoretical model will be acceptable if the conditions for its reliable use are satisfied or, even if there is doubt over this, it is accepted for reasons of economy or practical necessity in the context of the dispute. Mostly, proof of delay is a matter of fact but sometimes facts alone cannot answer the question and the law is required to take a position.

Irrespective of the method or approach used in analysing delays, what needs to be considered are the actual facts on-site to determine whether a particular event delayed the completion.

The various delay analysis approaches in use are (Society of Construction Law 2017).

Windows Analysis (or Time Slice Analysis)

Window analysis is based on the analysis of the effects of delay events over the entire length of a project by examining the events which have affected progress within 'windows' of the contract period sequentially. The duration of each window is not predetermined, but is frequently taken as being one month. At the end of each window the as-planned programme is updated to take account of any delaying inefficiency which is the contractor's risks, any necessary logic or duration revisions because of mitigation measures undertaken, and all excusable and/or compensable events during the period since the last update. The closing of a window in this way

[16]This falls under 'concurrency'.

[17]See Chapter 7 for further reading.

forms an as-built programme, at the end of that window, which effectively becomes the as-planned programme for the next window in sequence.

Collapsed As-Built Analysis

This method involves removing, from the as-built programme identified, excusable delays to show what the completion date would have been if those delay events had not occurred.

Impacted As-Planned Analysis

This method adds an identified excusable delay event (or events), either as a separate activity (or activities), or onto the duration of an existing activity (or activities), into the as-planned programme. The duration of the activity is derived (where possible) from the resource allowances on the as-planned programme. The impact of this added activity on the projected completion date is then compared to the original completion date to assess the extension of time due date.

Delay Costs Quantification

Delay claims are claims for additional time-related costs associated with the delays caused. Therefore, a delay claim is a claim for delay costs where a contractor must show that the cause of the delay is one that entitles the contractor to payment for the extra costs incurred.

To establish an entitlement to delay costs, a contractor must demonstrate that a delay to completion of the contract has occurred and must show that the cause of the delay is one that provides the contractor with an entitlement to extra payment, either under a term of the contract or for breach of contract by the employer. The claim must be supported by evidence of the facts on which it is based.

To warrant the payment of delay costs, a delay must affect the critical path and delay completion either of the whole of the works or of a milestone. If claims for delay costs are not handled appropriately there are risks that excessive costs will be incurred or that contract disputes will occur.

To quantify a delay claim, a contractor must demonstrate that the delay caused damages in the case of a breach of contract, or extra cost in the case of a specific contract provision. The onus is on the contractor to prove that the costs claimed have been incurred and that every effort has been taken by the contractor to minimise these costs. If a contract does not include a prescribed delay cost rate, then it is necessary to assess what delay costs are legitimate and to evaluate those costs.

On projects where the level of construction activity varies significantly between the various stages of construction, the appropriate costs will be those which relate to the periods in which the delay has occurred. The appropriate rate for calculation of site overheads is that related to the sequence of activities which were delayed, together with any consequential effects attributable to the delay.[18]

Site overheads are broadly related to the direct cost of the works to be undertaken. For projects with a high labour content, the cost of supervision will be greater than for projects with a high plant or material content. Less supervision is required for works where subcontractors are used compared with works requiring unskilled labour or using the contractor's employees.[19]

One method of assessing on-site delay costs is to evaluate the actual costs incurred. If cost information is provided by a contractor to justify a claim, then this must be audited to eliminate all costs that should be included in the direct costs of construction activities.

Generally, there will be very few material costs to site overheads. Costs that are not time-related would be deducted. For example, mobilisation and removal costs will be incurred irrespective of the contract duration. The exception could be where equipment or staff may be demobilised temporarily at the beginning of a long delay and remobilised at the end of the delay period in order to save the cost of retaining these resources on the site for the duration of the delay period.

The actual cost of staff and labour will generally be provided by wages or salary sheets and the cost of all external plant and services will be substantiated by invoices.[20]

The importance of site records is to assist in recognising that excessive costs cannot be overstated. Where the contractor has maintained good site records, the checking of the actual times claimed is ideal in establishing costs and the best way to accurately assessing the costs. A contractor is required to mitigate costs in the event of delay. Excessive expenditure due to poor management or inefficiency, if proven, is not recoverable.

Disruption Claims

Loss of Productivity (Disruption v. Delay)

Disruption is often treated by the construction industry as if it were the same thing as delay. The two issues are, however, entirely separate matters. Delay is lateness,

---

[18]Some non-critical activities will be delayed by a delay to critical activities on which they were dependent. They may still be non-critical, but they will be undertaken at a later time. This shift in time may not result in additional site overheads overall. It may merely cause the overhead costs to be incurred at a later time.

[19]The cost of engineering staff on civil engineering or complex multi-disciplinary projects may be higher than on building contracts.

[20]The delay cost related to equipment is the equipment rental charge only. In some cases, there will be reduced hire rates or even no charge for standby or non-operational periods. Where contractor owned plant is involved, invoices are not likely to be available, so an analysis of the costs claimed will have to be made separately.

whereas disruption is loss of productivity, disturbance, or hindrance or interruption to the contractor's normal working methods, resulting in lower efficiency (Gibson 2015).

Disruption costs may be distinguished from delay costs by virtue of the fact that the latter are a function of time and the former are essentially productivity-related. In a disruption claim, contractors claim that they could not achieve their planned output, because of the employer's actions or other causes not their responsibility, and hence that the damages or extra costs are payable.

Disruption, also called loss of productivity, results in a disruption to the work being carried out and not necessarily in a delay to the completion of the works. The works produced are not changed; it simply took longer to complete them because of the disruptions. A contractor may claim disruption costs independently of an extension of time claim, especially where the contract makes provision for this.

Disruption compensation is only recoverable to the extent that the employer caused the disruption. Most standard forms of contracts do not deal expressly with disruption; therefore, disruption may be claimed as a breach of the term generally implied into construction contracts—essentially, that the employer will not prevent or hinder the contractor in the execution of its work.

Causes of disruption can be broken down into either external or internal causes. External causes of disruption are generally not related to the project itself and will often fall into the *force majeure* category. These causes will also include government acts such as the passing of new regulations, changes to taxes, and new laws and regulations. Such events will generally involve certain rights of compensation to the contractor.

The internal causes of disruption can be causally attributed to the project itself, its planning and design and the manner in which the works are performed. Internal causes of disruption can be further broken down into:

1. Technical causes including changes in design, design errors and construction hindrance;
2. Difficulties in procuring materials, labour or skills;
3. Financial causes such as lack of employer funds; Material, labour cost increases and interest rate rises.

It will usually be expected on a complex project that a certain amount of uncertainty and rework will be expected at various stages of construction. Even when the project is going well, normal disruptions, made by both the contractor and the employer, will involve a certain amount of rescheduling and planning and additional costs to rectify. Despite the fact that these costs are built into the initial tender and will be absorbed without affecting the time frame or budget, it is possible to drastically underestimate the costs of such factors (Gibson 2015).

Disruption Substantiation

A disruption claim must identify specific events that are breaches of contract by the employer or events for which the contract specifically provides for extra costs. An example of disruption is where the employer ordered a contractor to cease work on a particular activity for frequent short periods (for example to provide for some necessary operational function of a plant or building) and the need for such stoppages was not specified in the tender documents. Another example would be where the employer ordered urgent variations and the contractor's labour force had to continually move from one activity to another at short notice, being unable to develop optimum productivity on a particular work activity.

To justify a disruption claim, a contractor must establish that the actual progress of the works has been interrupted and that the cause of the disruption was either a breach of the contract by the employer or an action for which the contract provides for the reimbursement of extra cost. Even if the employer does cause disruption, this may not result in an entitlement to additional payment. It may be that the contractor failed to comply with certain contractual requirements and is, therefore, not entitled to reimbursement of the disruption costs.

A standard normal requirement for the validity of a disruption claim is that the contractor must give notice of information required at the time of knowing of the disruption. The contract usually will provide the mechanism for the notification of the type of information required and to whom the notice should be addressed.

A disruption claim, in particular, may be difficult to ascertain due to lack of records and the difficulty of establishing the nexus of cause and effect. Just because it is difficult to establish such a claim, however, it does not mean that a contractor should not submit one. Equally, it does not mean that the contractor should not make a reasonable assessment of the damages incurred (see the section of this chapter regarding global claims).

Ideally, it should be established that before a particular event occurred, productive resources were achieving a certain level of productivity/financial income for the contractor. After that particular event occurred, however, and because of it, the same level of productivity/financial income could not be achieved.

There are several types of approach in respect of a disruption claim that may be applicable, depending on the nature of the work, the circumstances and the records available. The main approaches used must:

- show that resources were standing idle when they should otherwise have been operational—thus meaning that it took longer to do the same work;
- illustrate how the work of a particular trade has been prolonged, and apply a disruption factor in the same proportions to the cost of the trade;
- estimate a disruption factor; and
- illustrate the resources employed in producing specific and evidenced amounts of work prior to an event, and claim the extra resources doing equivalent work

after the event; Show that more expensive resources were required to do the same value of work because of an event, even though the level of resources, e.g. labour and plant hire, was the same (Gibson 2015).

Disruption Costs Quantification

A contractor must quantify the disruption costs once it has established that loss of productivity has occurred and caused a delay. This involves comparing the actual cost with what the cost would have been had it not been for the disruption. The contractor must demonstrate that the latter cost is reasonable, although it is hypothetical to some extent. In making such claims, a contractor must also establish that everything reasonable has been done to minimise the cost of the disruption. Cost details of the affected resource may then be compiled from the site accounting records. On this basis any resulting disruption claim will be in respect of actual loss and expense incurred instead of in reference to tender allowances.

The practices, which are determinant to success and failure of entitlement to contractor compensation due to disruption, are summarised as follows:

1. The work that has been affected must be clearly identified, and the work activities that were affected by the disruption must be specified. The extra expense incurred must be explained.
2. The contractor must show that the event leading to the disruption and financial loss was either a breach of contract, or an event provided for in the contract for which the employer is to be made financially liable to the contractor.
3. It must be shown that the actual work progress has been negatively impacted. It is not sufficient to show that planned future work has been impacted, as such uncertainties may never materialise.
4. The contractor must quantify the disruption costs using a selected and approved method of quantification. The principle that applies is that the extra costs incurred, compared to the costs that would have occurred had the disruption not occurred, are recoverable by the contractor.
5. The contractor sets out what the actual costs would have been had the disruption not occurred. This provides the base line to be used in the calculation.
6. The contractor must show that he has taken all reasonable steps to mitigate his loss, such as returning leased equipment, working on other parts of the project that were not affected by the disruption, and redeploying expensive resources so that they are not unnecessarily sitting idle.

Costing in Construction

In most cases, a contract will be awarded after a competitive bidding process, whereby the contractor awarded the job will be the one who gave the lowest bid for the works. This will generally be the result of extremely optimistic calculations that have been made by the contractor; calculations that have allowed the contractor to arrive at a price for the job that is lower than more moderate or conservative calculations made by the unsuccessful bidders.

Any contractor who factors in all of the risks and makes contingencies for some of the uncertainties facing the project, will rarely come in at the lowest bid. In fact, it is reasonable to conclude that in the case of the lowest bid winning a contract, it is not the contractor that is best placed to perform the task on time and on budget who will be awarded the contract, but rather, it will be the contractor most desperate for the works, and therefore most willing to make the unrealistic promise without planning for difficulties that could arise. Therefore it is wise to foster most of the issues in claims where they are brewing even before the contractor steps on-site (Burr 2016).

A rational system for the determination of the price for construction works might involve three principal elements:

1. The tendered price for which the contractor is willing to do the work.
2. Some method of assessing the suitability of that price, by way of a breakdown; A method or schedule for pricing any additional works or changes to be made to the scope within the tendered price.

Major projects for engineering works have traditionally been so uncertain in their scope, by definition, that the new contracts are drafted on the basis that they will not be tendered for a fixed price. Rather, the payment system is based on a remeasurement basis. Some are even procured on a cost-plus basis.[21]

Furthermore, the traditional payment systems have the characteristic of postponing the resolution of uncertainties until the later stages, or even after completion, of the project.

Direct Costs

These costs are sometimes known as ‘preliminary costs’, but are most often referred to as ‘site overhead costs’. They must be distinguished from the direct costs of construction activities. Each construction project entails certain indirect expenses that are charged directly to the job. These include, typically, costs of supervision, rental and depreciation of site offices, plant and machinery, site services, insurance and bond premiums.

When these costs are incurred as a result of delay and disruption for which a contractor is entitled to reimbursement, the contractor is usually entitled to recover

[21] See Chapter 3 for details.

an amount to cover these costs. To minimise the administrative effort required to provide actual cost information in support of the direct costs in a delay and disruption claim, a contractor will often provide estimated daily or weekly costs for the staff, plant and facilities involved. If good site records exist, it is usually possible to check such claims in broad terms and highlight inconsistencies in a contractor's claims. The onus is on the contractor to prove the amount of such claims. An examination by both parties of cost records may become essential.

Site-direct costs cover the costs of the items a contractor must provide during construction, including:

1. Salaries of site supervisory staff, including accommodation and travelling expenses.
2. Labour engaged on a part-time basis on these activities.
3. Skilled labour wages and overtime expenses.
4. General construction plant.
5. Major temporary works such as dewatering equipment, water supply, shoring and underpinning.
6. Small tools and consumables.
7. Site supply services including power, water, air and telephones.
8. Site offices, amenities, workshops and stores.
9. Site office expenses including couriers, postage, telephones and copying.
10. Insurance, security charges and long-service levy charges.

Indirect Costs

Costs claimed as indirect costs may in reality be contract-related services that are performed away from the site. These works relate to procurement, shop drawings, administration and accounting. In a project requiring design and build, the design works, issue of drawings and revision of plans can constitute a large part of the indirect costs.

The following types of cost may be considered in the category of offsite overheads:

1. Executive and clerical salaries.
2. Office occupancy costs (rent, mortgage, services, etc.).
3. Design office overheads and testing facilities.
4. Plant workshops, yards and storage areas.
5. General maintenance and depreciation of plant.
6. Advertising, marketing and general administrative costs.
7. Professional fees.
8. Offsite vehicle expenses, office supplies and taxes.

Labour and Equipment Costs

It may be necessary to redeploy resources from other contracts or engage subcontractors at higher rates. Redeployment of labour and sometimes equipment from one part of the site to another may also cause secondary disruption. The test for inclusion of labour and equipment in a delay and disruption claim is whether or not the contractor was present on the site and idle for a longer time due to the delay; or whether the contractor is one who is engaged on a specific work activity and whose time on-site is governed by the rate of progress of that activity.

Subcontractors' Costs

These include nominated, designated and selected subcontract delay costs. Subcontractors may well have an entitlement to delay costs from a contractor. In turn the contractor may consider that such delays were caused by the employer, and forward these claims to the employer.

Subcontractors' delay claims may include delay costs due to the actions of the contractor and the employer. The employer must ensure that these claims are separated and must insist that the contractor's claim does so. Contractors are often reluctant to provide such a break-up and endeavour to recoup all of a subcontractor's delay costs from the employer.

It often happens that in the subcontract there is no specified programme and the subcontractor has agreed to carry out the subcontract work in accordance with the requirements of the progress of work under the main contract. In that event it may be extremely difficult to prove that a particular act caused delay, since there would be no baseline programme by which to measure delay. To evaluate such claims, it is necessary to have the contractor establish the justification and amount of any claim in the same manner that would be required for the contractor's own claims; use a copy of the subcontracts between the contractor and the subcontractors to check what claims are justified; and, importantly, to have the contractor either certify that the subcontractor has been paid the entitlement due or else give a direction for the employer to pay the subcontractor.

Financing Costs

This is an area where claims have little or no justification. Sometimes, however, a contractor can genuinely incur an extra cost for financing charges. Note that if some delay costs, such as payments to suppliers, have not been paid out by the main contractor at the date of a payment claim, overdraft interest could not have been properly included in the claim.

If financing charges are claimed, the contractor should be required to provide proof of the dates upon which amounts were allegedly paid out. Bank statements with payments identified may be of assistance. Proof of payment of interest by the contractor should also be requested. The account may be in overdraft and it may only be necessary to have proof of the overdraft rate. Interest from the date when the money was paid out until the next date thereafter for a payment claim is then calculated.

Reasonable Time

Where the contract has no express provisions as to time for completion of the works, the contractor is obliged to complete the works within a reasonable time. What is reasonable is primarily a question of fact and depends upon all the circumstances that might be expected to affect the progress of the works: both what was anticipated at the outset, including the anticipated level of resources, and events which may occur during the project and over which the contractor had no control.

In calculating a reasonable time, all the applicable circumstances should be taken into consideration, such as the nature of the works to be done, the time necessary to do the works, the ability of the contractor to perform and the time that a reasonably diligent contractor would take to perform a similar task with similar constraints.[22]

Therefore, reasonable time is primarily a question of fact, and it must depend on all the circumstances that might be expected to affect the progress of the works. What constitutes a reasonable time has to be considered in relation to the circumstances that existed at the time when the contract obligations were performed, but excluding circumstances that were under the control of the contractor.

Obviously, to allow a contractor to complete in a ‘reasonable time’—a period that would be difficult to determine—would not be satisfactory to most employers and, because of this, the normal position is that a contractor has an obligation, as set out in the contract, to complete a project in a certain time period or by a certain completion date.

Of course, the date for completion as set out in the contract would cease to apply if there was a delay caused by a breach of contract or an act of prevention by the employer that prevents the contractor from completing the works by the completion date.

[22] *British Steel Corporation v Cleveland Bridge & Eng Co. [*1981*] 24 BLR 100.*

In such a situation, and in the normal course of events, the obligation to complete no longer relates to a fixed completion date but reverts to being an obligation to complete within a reasonable time. A consequence of there being no applicable fixed date for completion of the works is that any liquidated damages provision within the contract becomes ineffective (as there is no fixed date against which the liquidated damages amount can be applied), and the employer's entitlement to damages, which would only apply for a delay beyond what may be determined as a reasonable time period for completion (which may itself be very difficult to establish), would be financial compensation for the losses that the employer could actually prove that he had incurred.[23]

Because of the obvious difficulties, as outlined above, of allowing a contractor to simply complete in a reasonable time, most construction contracts provide for extensions of time to the completion date to be granted to the contractor because of particular events.

Time of Essence

The general rule is that time is not of the essence unless the contract expressly so provides. 'Time is of the essence' means that performance by one party at or within the period specified in the contract is necessary to enable that party to require performance by the other party. Therefore, any delay, reasonable or not, slight or not, will be grounds for cancelling the contract (Haidar 2011).

There is no general concept of time being of the essence of a contract as a whole. Instead, the question is whether time is of essence of an individual term. Time is not being considered to be of the essence unless the parties expressly stipulate that conditions as to time must be strictly complied with and the nature of the subject matter of the contract or the surrounding circumstances show that time should be considered to be of the essence.

Time of the essence clauses generally carry far less weight in construction contracts, as most construction contracts incorporate a variety of terms compelling the contractor to perform its duties in a timely fashion, such as liquidated damages and express termination provisions specifically addressing delays in performance. These specific clauses may well override a generic clause declaring time to be of the essence as they raise the question as to whether the parties intended the clause to operate in a field occupied by an express provision. Indeed, there is a good argument that where a party stipulates for liquidated damages, it has declared an intention that damages are an adequate remedy, meaning the time obligation is not a condition that would entitle that party to terminate the contract.

Another difficulty, in giving effect to a time is of the essence clause in construction contracts, is the sheer number of time references in construction contracts for various duties, obligations and notices. A missed time deadline in a construction contract may

[23] See Chapter 7 for further reading.

well arise after substantial performance leading to concerns of unjust enrichment. Accordingly, there may be reluctance in arriving at an interpretation that permits termination.

In *Mount Charlotte Investments Ltd v Westbourne Building Society*,[24] J. Templeman sets, *obiter dicta*, the three conditions that make time of essence as follows:

- the contract expressly stating that this is so;
- implication because of the special matter (for example the completion of the project will allow the commencement of another); and
- notice from the innocent party making time of the essence after the other party has defaulted under the clause.

In view of this, the insertion of a clause declaring time to be of the essence in a construction contract, unlike its insertion in most other contract forms, will not normally, in and of itself, allow the innocent party to rescind or terminate the contract for any breach of a time condition. In determining the party's intentions, the court will look to all the particular terms and circumstances and may very well import little meaning to the time is of the essence clause.

Global Claims

A global claim, also called a 'total loss claim', is a claim for financial loss which arises from various different events; however, the requirement to link cause and effect is largely absent (i.e. individual sums of money are not claimed for each individual event). Instead a single global sum is claimed in respect of the alleged cumulative effect of all of the events.

Unsurprisingly, global claims have been the subject of much controversy over the years and have been considered in many court cases, and the whole area remains a difficult and developing area of law (Haidar 2011).

The general rules that appear to being developed in respect of global claims are:

1. In most cases, individual causal links must be demonstrated between each of the events for which the employer is responsible, and particular items of loss and expense.
2. In circumstances where it is impossible to separate the specific loss and expense caused by a number of different events that are the responsibility of the employer, then these can be pleaded as producing a cumulative effect. In these circumstances, it is not necessary to break down each event and isolate the loss caused by each.
3. Where it is shown that some of these events (albeit not a 'significant' amount in causal terms) are not actually the responsibility of the employer, the global claim

[24][1976] 1 All ER 890.

need not necessarily fail, since it may be possible for the judge, or arbitrator, to apportion the loss between the causes for which the employer is responsible and other causes.

4. When pleading the claim, the particular events and heads of loss should be set out in reasonable detail. There will, however, usually be no need in commercial cases to do more than simply to plead the proposition that the particular events caused the relevant heads of loss.

In the case *Bernhard's Rugby Landscapes Ltd v Stockley Park Consortium Ltd,*[25] the courts considered all the major cases concerning global claims and as a result produced a summary of the position (at that time).

That position was that:

- While a court will approach a global claim or a total cost claim with caution, such claims are not necessarily bad and, in some circumstances, it may be the only way that a claimant can establish its loss.
- A global claim is permissible where it is impractical to disentangle that part of the loss attributable to each head of claim.
- The power of the court to strike out is very limited and should only be used where a claim is so evidently untenable that it would be a waste of resources for this to be demonstrated only after a trial.
- The fundamental concern of the court is that the dispute between the parties should be determined expeditiously and economically, and above all fairly.

Global claims were reviewed in the *Walter Lilly & Company Ltd v Mackay & Anor*[26] case. In that case, Mr. Justice Akenhead concluded that in relation to "global' or 'total' cost cases, claims by contractors for delay or disruption related to loss and expense must be proved as a matter of fact and the contractor needs to demonstrate on a balance of probabilities that, firstly, events occurred which entitled it to loss and expense; secondly, that those events caused delay and/or disruption, and thirdly, that such delay or disruption caused it to incur loss and/or expense.[27]

Several cases decided in the USA accepted a variant of the global claim approach, and this approach, often referred to as the 'modified total cost method', has been given approval in *Dillingham-Ray Wilson v City of Los Angeles.*[28] The modified total cost method addresses the shortfalls of the total cost method approach,[29] by segregating the aspects of the work affected by the project owner's shortcomings. Further, the

[25](QBD 1997) 82 BLR 81.

[26][2012] EWHC 1773.

[27]He further emphasised that it is open to contractors to prove these three elements with whatever evidence will satisfy the tribunal and the requisite standard of proof.

[28]*(2010) 182 Cal. App . 4th 1396.*

[29]The total cost method, similar to the global claim approach, is a methodology which compares the bid price of the project work with the total cost of the project work as performed, attributing the entire cost increase to the acts or omissions of the employer. Although advantageous to contractors, courts (as well as arbitrators and administrative boards) consistently have declared the method imprecise, and frequently have rejected it.

modified approach does not blindly accept the contractor's pre-bid cost estimates or the contractor's as-performed job cost records.

Acceleration

When delays occur and where the employer resists granting the contractor an extension of time that he is entitled to for an excusable event, or where an extension of time has been granted by the employer, but for a shorter period than the contractor is entitled to, a contractor may feel compelled to accelerate the works in order to overrun the completion date set by the employer, thereby avoiding exposure to liquidated damages. This creates an intricate situation in terms of the duty of care imposed on the employer and his representatives, and in terms of how the courts review such cases in providing relief for the contractor.

Acceleration disputes are also fairly common in construction. They arise generally when the employer instructs the contractor to complete the works within a shorter period than originally required under the contract (referred to as express acceleration), or when, because of one or more events for which the employer is responsible, the contractor must perform more work or delayed work within the same period (referred to as implied acceleration).

Acceleration can affect a contractor's costs in a variety of ways. If the contract provides for acceleration, payment should be based on the terms of the contract. If the contract makes no provision, the parties should agree the basis of payment before acceleration is commenced. The costs attributed to acceleration are:

- Additional labour and equipment costs arising from reduced efficiency of the expanded labour force and supplied equipment.
- Additional delivery charges for material and equipment required at the site outside of normal work hours.
- Costs of additional site facilities.
- Additional costs from advancing the date of delivery of manufactured elements.
- Overtime charges for operatives, engineers, staff and foremen.

Documentation and Records

While the documentation and recording of contract data and information is primarily for management purposes, it also places the contractor in a good position if he/she needs to pursue a claim. Those failing to collect and document information regarding changes and delay events, however, will be left with few options. Attempting to collect and assemble data after the event will naturally limit claims to retrospective accounts and analysis.

It is to be noted that the courts place a particularly heavy burden upon contractors in terms of the maintenance and presentation of documentation in support of any claim. The contractor has to maintain accurate and complete records, and should be able to establish the causal link between the client risk event and the resultant delay or disruption caused, and/or the loss and/or expense suffered (Haidar 2011).

Most claims, whether for time or money, involve establishing what is often called the nexus of cause and effect (i.e. the link between cause and effect). This means that what needs to be proved is that, because of the occurrence of a particular event, certain things happened, and as a direct result of those events happening this in turn has led to one of the parties incurring delays or costs which had not been previously contemplated by the parties and which it would not have been reasonable for the parties to have contemplated.

In order to establish this nexus (or link), which is never an easy task in construction, records need to be available which show that the circumstances that existed before an event occurred changed after that event occurred, and that that change in circumstances could only have been as a result of the event in question.

Project records may be as diverse as site investigation reports, feasibility studies, specifications, drawings, tender submissions, estimating and pricing details, minutes of meetings, formal instructions, test data, payment applications and certificates, weather records and so on.

The primary objective of the contractor at the outset of the construction contract should be to ensure that the appropriate and necessary procedures, records, documentation and correspondence are established and maintained throughout the entirety of the contract so as to ultimately facilitate successful completion of the contract and to avoid delay and disruption and other claims ending in dispute.

An equally important objective is to ensure that high standards of record keeping and documentation are maintained throughout the period of the contract to record the effect of delays, variations and other events, and that procedures in respect of the same are established and fully communicated to each of the contractor's relevant personnel involved in the contract.[30]

Records and information most likely to be crucial in the success of claims include[31]:

1. Master programme identifying the critical path and indicating how the contractor had envisaged the sequence and timing of the various activities based on the tender information.
2. Progress schedule of activities against the master programme.

[30]Many delay and disruption disputes could be avoided if the parties properly monitored and recorded the aforementioned information. Experts who advise on disputes often find that there is a lack of records, resulting in uncertainty as to when a delay occurred, who caused the delay, and the effects of that delay. Good recordkeeping can remove this uncertainty.

[31]The reality is that a small proportion of time, money and effort expended by the contractor in putting in place the aforementioned procedures and record keeping at the outset of a contract could ultimately save him a significant amount of time, effort and money at the end of the contract in having to recover, in arbitration or court proceedings, loss and expense incurred due to delay and disruption to the contract.

3. Estimate of weekly resources and anticipated expenditure to comply with the master programme.
4. Records of actual resources and expenditure based on progress.
5. Records of plant standing or uneconomically employed.
6. Labour allocation sheets and associated costs.
7. Variations register.

Mitigation

The contractor has a general duty to mitigate the effect on its works of the employer's risk events. This duty to mitigate does not extend to requiring the contractor to add extra resources, or to work outside its planned working hours, in order to reduce the effect of an employer's risk event, unless the employer agrees to compensate the contractor for the costs of such mitigation (Burr 2016).

In other words, mitigation measures are not the same as acceleration measures. It can be argued, however, that the obligation to progress the works diligently may require the contractor to take some positive action, and a failure to do so may result in liquidated damages being applied for the additional period of overrun which could have been avoided but for the failure to take action.

The contractor may react to the delays by making changes to his methods of working or sequence of working, or he may even accelerate the works. The issue then is whether he is entitled to recover the loss incurred by this reaction. The answer to this is that it depends on whether or not he was right to react as he did.

It is suggested that subject to the express terms of the contract, the contractor has no obligation or right to accelerate and is not entitled to recover additional costs incurred in acceleration measures to mitigate the effect of qualifying delays without an instruction from the client.

Examples of actions that could be taken by contractors to minimise delays and impact on the overall completion of a project and to mitigate losses include:

- Terminating hire of plant not being used, or hire out plant.
- Laying off workers who are not productive, when this is possible.
- Adding additional workers or subcontractors on activities impacted by delays.
- Working overtime.
- Reorganising work programmes and order of works.

Table 6.1 Loss and expense components

DELAY PERIOD			**0 Days**	
A - DELAY/TIME RELATED COSTS DUE TO THE CONTRACT PROLONGATION				
	Description	USD/Month	USD/Day	Total
A.1.	EQUIPMENT		0.00	0.00
A.2.	INDIRECT LABOR COST		0.00	0.00
A.3.	GENERAL EXPENSES		0.00	0.00
	TOTAL	0.00	0.00	0.00
A.1. EQUIPMENT				
	Equipment cost	USD/Month	USD/Day	
1.	Equipment Rental Cost		0.00	
2.	Equipment Spare Parts & Cons.		0.00	
3.	Vehicle & Buses Rental Cost		0.00	
4.	Vehicle & Buses Spare Parts & Cons.		0.00	
5.	Equipment rented from others		0.00	
6.	Fuel, Oil & Lube		0.00	
	TOTAL	0.00	0.00	
A.2. INDIRECT LABOR COSTS				
	Category	USD/Month	USD/Day	
1.	Management & Supervision		0.00	
2.	Engineering & Technical		0.00	
3.	Administration Staff		0.00	
4.	Quality Control		0.00	
5.	Part-time staff		0.00	
6.	Laboratory Staff		0.00	
7.	Medical Staff		0.00	
8.	Security Staff		0.00	
9.	Safety Staff		0.00	
10.	Operators & Drivers		0.00	
11.	Management Consultant		0.00	
12.	Audit		0.00	
13.	Temp. Facilities O&M Staff		0.00	
14.	Warehouse Staff		0.00	
15.	Workshop Staff		0.00	
16.	Cooking Staff		0.00	
17.	Client support services (labour only)		0.00	
	TOTAL	0.00	0.00	

(continued)

Table 6.1 (continued)

A.3. GENERAL EXPENSES

	Category	USD/Month	USD/Day
1.	Insurance Cost		0.00
2.	Bank Guarantee		0.00
3.	Income Tax		0.00
4.	Consultant office maintenance		0.00
5.	Communication		0.00
6.	Office Supplies & Stationery		0.00
7.	Client Support Services		0.00
8.	Social Insurance		0.00
9.	Service Award		0.00
10.	Food & Water & Electricity		0.00
11.	Housing Indirect staff		0.00
12.	Staff Transportation		0.00
13.	Travel Expenses		0.00
14.	Business Travel		0.00
15.	Medical Expenses		0.00
16.	Sick Leave Pay		0.00
17.	Cleaning Up		0.00
18.	Visa & Iqama Fees		0.00
19.	Licences & Permits fees		0.00
20.	Safety & Protection		0.00
21.	Small tools & Construction Consumables		0.00
22.	Site facilities maintenance including camps		0.00
23.	Temporary facilities Depreciation costs		0.00
	TOTAL	0.00	0.00

A.3.1. INSURANCE COST

	Category	USD/Month	USD/Day
1.	Contractor's All risk Ins including Workman compensation Policy		0.00
2.	Equipment & Vehicle Insurance		0.00
	Total Insurance Premium Cost for one Month	0.00	0.00

(continued)

Table 6.1 (continued)

A.3.2. BANK GUARANTEE COST			
	Category	USD/Month	USD/Day
1.	Performance Bond		0.00
2.	Advance Payment Bond		0.00
	Total Bank Guarantee Cost for one Month	0.00	0.00

A.3.4. CONSULTANT OFFICE MAINTENANCE (USD/Month)	
1.	Stationery
2.	Kitchen Supplies
3.	Water & Electricity
4.	Telephone
5.	Maintenance
6.	Other Items (Cars etc.)
	TOTAL

A.3.23. Temporary facilities DEPRECIATION COST (USD/Month)			
1.	On Site Camp		0.00
2.	Site Offices		0.00
3.	Warehouses		0.00
4.	Temporary fencing & lighting		0.00
5.	Access Roads & Guard Houses		0.00
	Total	0.00	0.00

B. DISRUPTION		
Man-hours Loss on	Hours	
Man-hours Loss on	Hours	
Total Man-hour Loss		0.00
Unit Rate Per Hour	USD/Hour	
Disruption costs		0.00
C. ESCALATION		

D. FUNDING COSTS		
Description		Amount USD
Delay Costs		0.00
Disruption Cost		0.00
Escalation Costs		0.00
Funded Costs	USD	0.00
Recovery		
Funding Period	Days	0.00
Effective Period	Days	0

(continued)

Table 6.1 (continued)

Overdraft Rate =	xx%	
Funding Cost per Effective Period	USD	0.00
E. RETENTION FUNDING		
Description		Amount USD
Contract Price		
Retention	%	
Expected Retention		0.00
Recovery		
Funding Period	Days	0.00
Effective Period (50%)	Days	0
Overdraft Rate	12%	0.12
Retention Funding Cost	USD	0.00
F. CLAIM PREPARATION COST		
Prepation of Statement of Claim Cost		0.00

Description	**Cost USD/Month**	**Cost USD/Day**	**Total cost USD**
(A) Delay Cost	0.00	0.00	0.00
(B) Disruption Cost	–	–	0.00
(C) Escalation Cost	–	–	–
(D) Funding (Financing Cost)	–	–	0.00
(E) Retention (Financing Cost)	–	–	0.00
(F) Claim Preparation cost	–	–	0.00
(G) Liquidated Damages	–	–	–
(H) Engineer's Costs	–	–	–
(I) Other Costs	–	–	–
	–	–	–
TOTAL	**–**	**–**	**0.00**

References

Society of Construction Law (February 2017) Delay and Disruption Protocol, 2nd edn

Books

Burr A (2016) Delay and disruption in construction contracts. Routledge, Abingdon

Gibson R (2015) A practical guide to disruption and productivity loss on construction and engineering projects. Wiley-Blackwell, Hoboken, NJ

Haidar AD (2011) Global claims in construction. Springer Verlag, London

Haidar AD, Barnes P (2017) Delay and disruption claims in construction. ICE Publishing, London

Case Law

Bernhard's Rugby Landscapes Ltd v Stockley Park Consortium Ltd (QBD 1997) 82 BLR 81
British Steel Corporation v Cleveland Bridge & Eng Co. [1981] 24 BLR 100
Dillingham-Ray Wilson v City of Los Angeles(2010) 182 Cal. App. 4th 1396
John Barker Construction Limited v London Portman Hotel Limited (1996) 83 BLR 31
Mount Charlotte Investments Ltd v Westbourne Building Society [1976] 1 All ER 890
Walter Lilly & Company Ltd v Mackay & Anor[2012] EWHC 1773

Chapter 7
A–Z Guide to Contract Management and Law

Abstract This chapter addresses many important headings, conditions, topics and doctrines related to understanding contract management and construction contract law that contract managers and professionals in the construction industry must be familiar with. Each heading is discussed briefly in order to give the reader an understanding of the topic. The chapter provides a comprehensive guide to construction law's main topics by listing these topics in alphabetical order for ease. This methodology makes, as well, an ideal introduction to the subjects for those studying contract management in a professional capacity. There is no comprehensive review of each topic, but an introduction to familiarise the reader with the subject. It is certain that some topics deserve a chapter to themselves; however, in this book the purpose is to identify and introduce these headings to give the reader a sample of the wide range of topics that can be addressed and to provide an introduction and general knowledge of these topics.

Keywords Bargain · Bonds-guarantees—warranty · (The) engineer · Programme—schedule · Termination

Adjudication

Adjudication is a contractual or statutory procedure for swift interim dispute resolution. It is provided by a third-party adjudicator selected by the parties in dispute (Martinez et al. 2012).

Parties to a construction contract may be able to refer disputes to adjudication, a cheaper and quicker way of resolving issues. Adjudication is designed to offer a quick cash-flow remedy during the process of a construction project. For parties to refer a dispute to adjudication, the contract must have expressed terms to do so.[1]

[1] Adjudication is practical for dispute resolution for many sub-contractors and smaller construction companies as they are often unable to afford to dedicate the time and money to court proceedings or arbitration.

A. D. Haidar, *Handbook of Contract Management in Construction*,
https://doi.org/10.1007/978-3-030-72265-4_7

Disputes are in many occasions referred to an independent dispute adjudication board (DAB, or simply 'the board') as a precondition to arbitration.[2]

DAB is a standing board in the sense that it is formed at the signature of the contract and remains in place continuously until the works are completed. Typically, this means that the DAB is in place for a period of years, the exact length depending on the duration of the particular project.

It would be more appropriate (and, probably, less expensive for the parties) to provide in the general conditions for an ad hoc DAB; that is, a board which would only need to be constituted if and when a dispute or disputes should arise, and which would normally cease to operate once a decision on such a dispute or disputes had been issued.

The board's decision should set out briefly the matter in dispute, the relevant facts, the principles (including contractual provisions) to be applied, and the basis for its decision. Settlement of disputes by the DAB is as follows:

1. A party gives notice of its intention to refer a dispute to the board.
2. Within 14 days[3] thereafter, the parties must jointly appoint a board.
3. After a DAB has been appointed, a party may refer a dispute to the board for its decision.[4]
4. Where either party has given such a notice of dissatisfaction, the parties are required to attempt to settle the dispute amicably for an agreed period of days. After the expiration of this period (and assuming no amicable settlement), each party is free to initiate arbitration as to the specific dispute.

Balance of Probabilities

The object of any claim is to convince the party responsible for providing a decision or determination that the claimant has the right to be compensated in ether time or money for the event from which the claim arose. The claimant has the obligation to prove that his claim is just and the standard of proof required to do so is based upon the 'balance of probabilities'. If the claim does not fulfil these requirements, there is a very good chance that it will fail because the claimant has simply not adequately demonstrated his case.

Hence, the 'balance of probabilities' in construction relates to the standard of proof required to determine whether there are sufficient facts and documents to substantiate a claim.

[2]In many standard forms of contracts, administration of the contract is done by the engineer, who is required to act impartially. They also require that disputes be referred to the engineer for decision, as a precondition to arbitration. A dispute adjudication board is usually inserted to replace the engineer in determining disputes under those contracts.

[3]Or an agreed period.

[4]The DAB has (an agreed period) days in which to give its decision, which must be reasoned and is binding on the parties. If a party is dissatisfied with the decision it must give a notice of dissatisfaction to the other party within (an agreed period) days after receiving the decision.

All forums, including courts and tribunals, must apply the appropriate standard of proof when deciding if a claim has been proven. The overarching goal is for the court or tribunal to be satisfied to the requisite degree of the claim being submitted. In all construction matters, the decision maker must be satisfied that the matter has been proven on the balance of probabilities, while criminal matters require a higher standard of proof.

In construction claims, the law is highly sympathetic to the difficulties the claimant faces in proving the non-breach position. It relaxes the burden on the claimant. This approach appears to have started with the seventeenth-century case of Armory v Delamirie.[5] The Armory principle has been extended to the situation where the defendant's breach meant that it had failed to keep records it should have kept and which would have helped the claimant to prove its case, even if this was not a deliberate attempt to thwart the process of proof in court.

In the 1930s case of Briginshaw v Briginshaw,[6] two very important considerations were established in order to assist the court and individuals in determining what surpasses or ultimately fails to reach this standard of proof, namely that a court or tribunal must be satisfied of the matters before it on the balance of probabilities:

1. To a comfortable degree.
2. Based on very clear and cogent evidence.

It is this notion that guides the tribunals on how best to apply this threshold.

Bargain

It is not uncommon for a powerful employer to impose its 'written standard terms of business' upon the contractor, nominated subcontractors and the consultant, who traditionally possess considerably less bargaining strength (Beale 2020).

Where the use of tailored contracts or standard form of contracts is accompanied by inequality of bargaining power, there is a greater likelihood of their being used as instruments of economic oppression because their terms can more easily be weighted in favour of the interests of the stronger parties who prepare them. This can fall under *contra proferentum*, as described in Chap. 2.

There is no presumption by the courts that contracts are fair and reasonable and, as a result, they are more likely to be subjected to judicial regulation. In such circumstances, the courts may well take into account the absence of genuine agreement and justify their intervention on that basis.[7]

The test for understanding whether a contract is single sided and could legitimately be varied or annulled by the courts was set out as follows:

[5]*[1722] EWHC J94*, in which the defendant to a claim for trover of a jewel refused to produce it, thwarting the claimant chimney sweep's attempt to prove its value. In the circumstances, the court gave the claimant the benefit of any uncertainty.

[6]*(1938) 60 CLR 336.*

[7]*Balfour Beatty Regional Construction Ltd v Grove Developments Ltd v [2016] EWCA Civ 990.*

1. A has a contract with B for work.
2. Before it is done, A has reason to believe B may not be able to complete.
3. A promises B more time to finish.
4. A 'obtains in practice a benefit, or obviates a disbenefit' from giving the promise.
5. There must be no economic duress or fraud.[8]

It remains the case, even though some small contractors or subcontractors are likely to have little choice but to 'take them or leave them'.

Battle of Forms

The battle of the forms refers to a situation in which two companies, A and B, are negotiating the terms and conditions for a task or project whereby:

- A offers to buy goods and/or services from B on A's conditions.
- B accepts A's offer, and agrees to supply goods and/or services but only on B's own conditions.
- A acknowledges B's acceptance on A's conditions.[9]

This is the 'battle' and it continues for as long as A and B send each other orders, estimates, acknowledgements and other forms, each with A's or B's terms and conditions. Neither A or B ever agree to the other side's terms and do not consider whose terms are the most appropriate for the goods and/or services being provided.

Under the common law, it is a rule that the acceptance has to match the offer in every detail. This is known as the mirror image rule. If the terms of the acceptance differ from the offer at all, it constitutes a counteroffer. A counteroffer is a rejection of the original offer, plus a new offer, which vests the original offeror with the power to create a contract by accepting the counteroffer.

The traditional legal method of analysing what contract has been made as a result of the 'battles of the forms' concentrates on viewing each document as an offer, a counter-offer, or an acceptance. But the courts have clarified this point:

> In many of these cases our traditional analysis of offer, counter-offer, rejection, acceptance and so forth is out of date. The better way is to look at all the documents passing between the parties — and glean from them, or from the conduct of the parties, whether they have reached agreement on all material points — even though there may be differences between the forms and conditions printed on the back of them.[10]

The battle of the forms is an example of the problems which arise when a standard form of contracts is used by both parties to the deal. While standard forms of contracts can represent the intentions of the parties, it is clear that contractors, subcontractors

[8] *Williams v Roffey Bros & Nicholls (Contractors) Ltd [1989] EWCA Civ 5.*

[9] *Lord Denning in Butler Machine Tool v Ex-Cell-O [1977] EWCA Civ 9.*

[10] *Butler Machine Tool Co Ltd v Ex-Cell-O Corp (England) Ltd [1977] EWCA Civ 9.*

and even employers often proceed without reading the details embodied in a standard form.

Where you have a ‘battle of the forms’, the approach of the court will typically be thus (Beale 2020):

1. As in any other construction contract, the test is objective, albeit that the court must take into account the factual matrix—i.e. what actually happened.
2. In most cases, a contract is formed as soon as the last set of forms is sent and no objection is taken.
3. Acceptance by conduct can be inferred, although conduct will amount to acceptance only if it is clear that the party intended to accept the terms. Acceptance of a delivery, of itself, may not be enough.
4. Where the parties have not agreed which set of standard terms applies, then the only inference that can be drawn is that their agreement was made on the basis that neither set of standard terms would apply.
5. If neither party’s terms and conditions are to be incorporated into the contract this will usually mean that the contract will be subject to the terms and conditions implied by statute.
6. Whilst the subsequent conduct of the parties may be relevant to the enquiry as to whether particular terms were or were not agreed, the subsequent conduct of the parties cannot be relied upon as an aid to the construction of the contract.[11]

Bonds and Guarantees

Many construction projects today require that contractors provide bonds. Construction bonds are protection for the employer against non-payment, lack of performance, contracor default, and warranty issues.

The provision and maintenance of the Performance Security, the Advance Payment Bond (if applicable) and any other bonds shall be a condition precedent to the contractor’s entitlement to be paid and/or continue to be paid for performance of the works.

If, for any reason, any of the Performance Securities or the Advance Payment Bond referred to in the contract become invalid or the employer has reasonable grounds for believing that they are about to become invalid, or an insolvency event has occurred or is about to occur in relation to the bank providing the Performance Security or Advance Payment Bond, then the contractor shall, at the employer’s request, provide alternative bonds which are satisfactory to the employer within fourteen (14)[12] days of being requested so to do.

If the contract price is increased pursuant to the main contract, the contractor must provide additional security so that the increased amount of security is the percentage of the increased contract price.

---

[11] *Immingham Storage Company Ltd v Clear plc [2011] EWCA Civ 89.*

[12] Or as stated in the Contract.

In relation to any Advance Payment Bond, if the contractor has not repaid the advance payment in full, the contractor shall extend the validity of the Advance Payment Bond until the advance payment has been repaid in full.

As well, in relation to any Performance Bond, if the contractor has not become entitled to receive the performance certificate, the contractor shall extend the validity of the Performance Bond until the contractor becomes so entitled.

Advance Payment Bond

Where Advance Payment is to be made to the contractor by the employer pursuant to the contract price, payment terms and payment schedule, the contractor shall, on or before the date on which the contractor submits the first statement in which any amounts are claimed in respect of any advance payment, provide to the employer an Advance Payment Bond in respect of such advance payment. The contractor is required to provide an Advance Payment Bond for each advance payment under the following conditions:

1. It shall be in the approved form set out in the contract (Form of Advance Payment Bond).
2. It shall issue from a well-known and reputable bank located in the country where the works are being carried out, or for an approved established international bank.
3. It shall be for the amount that the relevant Advanced Payment is being made.
4. The provision and maintenance of the Advance Payment Bond shall be a condition precedent to the contractor's entitlement to be paid the advance payment.

Performance Security (Bond)

The contractor, in most contracts, must deliver to the employer a Performance Bond as Performance Security in relation to the works. The Performance Bond is referred to as the 'Performance Security' in many cases. Performance bonds are in place to guarantee that the project will be done according to the contract's specifications and on schedule.

If a contractor fails to follow contractual outlines by skimping on any part of the project, the bank, issuing the performance Bond, would be liable for paying up to the bond's face value.[13]

Any performance security required by the contract shall be provided:

[13] A simple example of a performance bond claim would be if a painter applied the finishing coat of paint but failed to apply primer first. In this instance the project owner could make a claim on the painter's performance bond if the painter refused to strip the paint and reapply it correctly, according to the contract's specifications.

1. Within seven (7) days[14] of signing the contract.
2. In the approved form set out in the contract (Form of Performance Bond clause).
3. From a well-known and reputable bank located in the country where the works are being carried out or for an approved established international bank.

Whether or not a Performance Bond is required will depend, in the main, on the perceived financial strength of the party bidding to win a contract, as the most common concern relates to a contractor becoming insolvent before completing the contract. Where this occurs the performance bond provides compensation guaranteed by a third party up to the amount of the performance bond.

Bonds are typically set at 5–10% of the contract value. This compensation can enable the employer to overcome difficulties that have been caused by non-performance of the contractor such as, for example, finding a new contractor to complete the works.

Bonds can be 'on demand' or 'conditional', with conditional bonds requiring that the employer provides evidence that the contractor has not performed their obligations under the contract and that they have suffered a loss as a consequence.

The obligation for the contractor to provide the employer with a Performance Bond is set out in the tender documents. From an employer's viewpoint it is wise to stipulate that the bond stays in place until the end of the defects liability period when the final certificate is issued.

Performance Bonds can be issued either by an insurance company or by a bank, and the cost of the bond is borne by the contractor.

Warranty Bond

A Warranty Bond is a legal document that guarantees to the employer that the contractor who carried out the work will fix defective works or materials should an issue arise during the warranty period specified in the contract.

The most important reason to work with contractors who acquire warranty bonds is the assurance of having recourse after the project is complete and accepted, should the employer find that there is a problem with the work or material installed on the job.

Bid Bond

The Bid Bond protects the project's employer if the bid is not honoured by the contractor. The employer is the obligee under the bond and has the right to sue the contractor and the bank or surety (the issuer of the bond) to enforce the bond.

---

[14]Or as stated in the contract.

If the contractor refuses to honour the bid, the contractor and surety or bank issuer of the bond are liable for any additional costs incurred in contracting a second time with a replacement contractor.

Breach of Contract

A breach of contract is committed when a party without lawful excuse fails or refuses to perform what is required from him under a contract.[15]

A breach of contract does not automatically bring a contract to an end. However, a breach of contract gives to the injured party a right to claim damages, and it may give him the additional right to terminate performance of the contract. The right to terminate performance of a contract may be by way of an express provision in the contract, or may be because the breach of the contract is of such a fundamental and repudiatory nature that it is considered to be a repudiatory breach of the contract or (if it is stated as being an intention in the future) an anticipatory repudiatory breach of the contract (Beale 2020).

Generally speaking, any breach of contract should fall into of the following:

1. Major contract breaches—also called material.[16]
2. Minor contract breaches—also called non-material.[17]

Generally the contract will set out what those breaches are. Some of the conditions that define a breach of contract by a contractor are:

1. Refusal to carry out work.
2. Abandoning the site.
3. Removing plant from the site.
4. Failure to make payments for his subcontractors.
5. Employing others to carry out the work.
6. Failure to allow access to the site to the engineer or employer and his representative.
7. Failure to proceed regularly and diligently.
8. Failure to remove or rectify defective works.

The potential remedies under a breach of contract are[18]:

1. Damages for breach of contract.

[15] *Bellgrove v Eldridge [1954] 90 CLR 613.*

[16] A 'material breach', as the term itself suggests, is the breach of a basic, main term of the contract, so primary that upon such a breach, the other reciprocal promises cannot be performed by the other party to the contract.

[17] A 'non-material breach' is usually considered to be less serious than a material breach of contract. A non-material breach, or 'minor breach', is one which pertains to an inconsequential or ancillary term that does not affect the whole outcome of the contract.

[18] *Hadley v Baxendale [1854] EWHC J70.*

2. Rescission or repudiation.
3. Specific performance.[19]

Thus, where a party gives the other party an immediate cause of action, it generally results in a right to damages as compensation for loss arising out of the breach. In other words, it is the violation of an obligation by one party, which by right accrues to the other party under the contract to obtain a remedy for the breach in an action for damages. This, however, does not relieve the injured party from the obligation to perform its part of the contract except when the breach goes to the root of the contract.

Burden of Proof

Ordinarily, the burden of proof lies with the party who makes an assertion, hence the legal maxim '*he who asserts must prove*'. Accordingly, it is not for a party defending a claim to disprove the claim; it is for the party pursuing a claim to prove the claim.

If, however, the party pursuing a claim adduces sufficient evidence to raise a presumption that what is claimed is true, the burden of proof will pass to the other party. It is then for that party to adduce sufficient evidence to rebut the presumption. Therefore, any claim document should be prepared with the aforementioned basic principle in mind.

The standard of proof required in civil proceedings is referred to as the 'balance of probabilities' principle, whereby the court/tribunal makes its decision on the basis that something is more likely to have occurred than not. This is to be contrasted with the standard of proof in criminal proceedings, which is set at a much higher level of 'beyond reasonable doubt'.

Following on from the above, the following initial points need to be proved in a loss and expense claim document:

- That an event actually occurred.
- That the event was one expressly catered for within the contract.
- That the notices required under the contract had been given.
- What the effect was, in financial terms, of the specified event.[20]

An example of how burden of proof is applied can be simplified thus:

- A contract is formed between two parties.
- The claimant alleges a breach of contract against the defendant, and claims damages.
- The claimant bears burden of proof to show that a contract was formed and damage suffered.

[19] *Critchlow, J. (2007) and Hochster v. De la Tour (1853) 2 El & Bl 678; 118 ER 922.*
[20] Hadley v Baxendale [1854] EWHC J70.

- The defendant admits the contract existed. The defendant alleges that the damage claimed comes within an exclusion of liability clause in the contract.
- The onus of proof is on the defendant to show that the damage falls within the exclusion clause.

Damages

The purpose of damages is to compensate the innocent party for the loss that he has suffered as a result of a breach of contract by the other party. The action for damages is always available, as of right, when a contract has been broken, other than where it has been excluded by an express term of a contract.[21]

The aim of an award of damages is to put the innocent party in the same position, as far as is possible by way of a payment of money, as it would have been in had the other party performed its obligations under the contract (i.e. if it had not breached the contract terms). The innocent party must take reasonable steps to mitigate his loss.

Damages cannot be recovered where the loss that the innocent party has suffered is too remote a consequence of the other party's breach of contract. The general principle is that the innocent party can only recover losses that were within the reasonable contemplation of the parties at the time that the contract was entered into.

The purpose of damages is to put the innocent party (i.e. normally the claimant), 'so far as money can do it' back to the same position as if the contract had been properly performed.[22]

The court will similarly take the claimant's overall position into account in determining the basis on which damages are to be assessed. It will not generally order the defendant to pay an amount that will actually make the claimant's position better than it would have been if the contract had been performed.

As a general rule, damages are based on loss to the claimant and not gain to the defendant; therefore, in general, punitive damages cannot be awarded in a purely contractual action, since the purpose of damages is to compensate the claimant and not to punish the defendant.

Sometimes the proper measure of damages is not the cost of reinstatement (which is more usual) but the difference in value between the work actually produced and the work that should have been produced. This will particularly be the case where the claimant has no prospect or intention of rebuilding, or where it would be unreasonable to award the cost of reinstatement.

---

[21] *Hadley v Baxendale [1854] EWHC J70.*

[22] *Whiten v Pilot Insurance Co. 2002 S.C.R. 595.*

Documents

The main documents that are needed generally in a project are:

- Contract Agreements
- Bonds
- Liability Insurance
- Bill of Quantities (BOQ)
- List of Materials
- Construction Schedule(s) (Programme)
- Drawings.

The above have been described one way or another in this book. This section, however, will look at other secondary documents that are usually carried out by the contractor.

Progress Reports

During the course of the works; the contractor has a duty to prepare and submit daily and weekly reports and monthly progress reports to the engineer.

Reporting shall continue until the contractor has completed all work that is known to be outstanding at the completion date stated in the taking over certificate to be issued for the works.

Each report should include, but not limited to, the following:

1. Charts and detailed descriptions of progress against the programme, including contractor's documents, procurement, manufacture, delivery to site, construction, erection, testing, commissioning and trial operation.
2. Photographs showing the status of manufacture and of progress on the site (all photographs will be labelled with the date on which they were taken, where they were taken and what it is that is shown).
3. For each main item of plant, goods and materials, the name of the manufacturer, manufacture location, percentage progress, and the actual or expected dates of commencement of manufacture.
4. Contractor's inspections; tests; and shipment and arrival at the site.
5. Copies of quality assurance documents, test results and certificates of materials.
6. List of variations, variation proposals requested by the engineer that are in the process of being prepared and/or considered, and notices for all claims.
7. Safety statistics, including details of any hazardous incidents and any activities relating to environmental aspects and public relations.
8. Comparisons of actual and planned progress, with details of any events or circumstances that may jeopardise the completion in accordance with the contract, and the measures being (or to be) adopted to overcome delays.

9. A detailed programme showing the timing and sequence of activities to be carried out over the three (3) month period following the month in which the monthly progress report is submitted.
10. The current contract price and the cost of the works already incurred and the current estimated cost of completing the works.
11. Notification of any claims which the contractor anticipates making in the next month.
12. A status report of any unresolved disputes.
13. Any outstanding information and/or approvals that the contractor anticipates requiring from the engineer or the employer over the next month.

Moreover, the contractor shall submit, to the engineer, details showing the number of each class of contractor's personnel and the number of each type of contractor's equipment on the site.

As-Built Documents

The contractor shall prepare and keep up-to-date a complete set of 'as-built' records of the execution of the works, showing the exact as-built locations, sizes and details of the works as executed, as set out in the specification and as may be required from time to time by the employer, any consultant, any relevant governmental authority and/or any relevant utility company.

Example of the types of as-built-documents are:

1. The depths of various elements of foundation in relation to ground floor datum.
2. Horizontal and vertical locations of underground services, utilities and appurtenances, referenced to permanent surface improvements.
3. The location of internal services, utilities and appurtenances and accessible features of the structure.
4. Site changes to dimensions and detail.
5. Changes made and/or confirmed by variation orders.
6. Any other matters specified as being required by the specification.

The contractor shall also ensure that specifications and addenda are clearly marked up to record:

1. The manufacturer, trade name, catalogue number and supplier of each item of plant and/or equipment installed.
2. Changes made and/or confirmed by variation orders.

Operation and Maintenance Manuals

Before commencing the tests prior to completion (save where the engineer instructs otherwise), the contractor shall supply to the engineer provisional operation and maintenance manuals in accordance with the specification and in sufficient detail for the engineer, the employer, any relevant governmental authority, utility company, operator or tenant to operate, maintain, dismantle, reassemble, adjust and repair the respective parts of the works or a section (as the case may be).

The form and numbers of such operation and maintenance manuals is submitted as set out in the specification and shall be provided in both English and any other specific language as set out in the contract.

The works or a section (as the case may be) shall not be considered to have achieved substantial completion for the purposes of being taken until the engineer has received and approved final operation and maintenance manuals in the required detail as set out in the specification, and any other manuals specified in the specifications for these purposes.

Economic Duress

Duress is a means by which a party may be released from the obligations under a contract where unlawful threats have been made.

The duress is of the sort that deprives the party of consent in entering the contractual arrangement, although the test at law is that the party is exposed to pressure and is deprived of choice that would otherwise be available. There must be effectively no choice other than to comply with the request or the demand to be successful in a claim for duress.

In the commercial context this vitiating factor may be alleged where illegitimate pressure has been made that would affect a party's economic interests. A typical example in construction is where the employer withholds payments for the contractor unless certain tasks are completed.

In construction, a contract entered into under economic duress is voidable and not void. A contractor or the employer who has entered into the contract may either affirm or avoid such contract after the duress has ceased; and if he has so voluntarily acted under it with the full knowledge of all the circumstances he may be held bound on the ground of ratification, or if, after escaping from this vitiating factor, he takes no steps to set aside the formed agreement he may be found to have affirmed it.[23]

Economic duress may apply to the formation of the contract, at the commencement of the performance of the contract or subsequent variations of the contract. The pressure must be of a nature that is illegitimate and that was a significant cause of inducing the party to agree to the terms of the contract. The threats made and

[23] *Occidental Worldwide Investment Corp. v. Skibs A/S Avanti, Skibs A/S Glarona, Skibs A/S Navalis (The 'Siboen' and the 'Sibotre') [1976] 1 Lloyd's Rep. 293.*

pressure asserted must have been particularly coercive and of some significant weight or gravitas. The injured party conduct must be affected in a significant way by the duress, and a reasonable alternative must not be available at the time of the duress (McKendrick 2020).

Economic duress is characterised by a lack of choice. Where an alternative is available to the injured party, the vitiating defence will not be available; however, the alternative must be reasonable. The following descriptive examples may give rise to a claim for economic duress:

1. Threats to terminate a contract, where the threat is properly regarded as illegitimate pressure.
2. Applying pressure in bad faith.
3. Making threats that are calculated to seriously damage another.
4. Threats to prosecute where the charge is known to be false.
5. Requirements for extra payments to be made over and above the original contract price.
6. Using knowledge of the affairs of the person suffering the duress to apply illegitimate pressure.[24]

The courts have generally held that the recovery of money on the ground that it had been paid under duress, other than under duress to the person, was not limited to cases where there had been duress to goods; the duress could also take the form of economic duress, which could be constituted by a threat to break a contract.

If, however, a party who had entered into a contract under economic duress later affirmed the contract, he was then bound by it.

The Engineer

For a typical construction project, the customary duties and role of the engineer are well understood. One of the most important roles of the engineer is his or her role during the construction process. Engineers and contractors understand the construction process, and their duties as set forth in their respective contracts, applicable laws and regulations (Poole 2016).

During the construction process it is critical that the employer get competent and professional engineering advice and guidance.

Without that advice, the employer will be limited in his or her ability to determine what he or she is entitled to receive from the contractor, and to accurately measure whether it has been received.

The engineer's role during the typical construction project is to administer the construction contract:

- Representing, advising, and consulting with the owner during the construction.

[24] *Barton v Armstrong [1976] AC 104.*

- Selecting/procuring and awarding a contract to a competent contractor at a fair price (subject to the vagaries of the competitive bidding process).
- Receiving and reviewing the contractor's submittals (usually referred to as shop drawings).
- Reviewing and approving the contractor's requests or applications for payment.
- Providing interpretations and clarifications of the requirements of the construction contract.
- Managing the changes to the requirements of the construction contact and to the design.
- Assisting the owner and contractor with resolution of disputes.
- Providing observations of the work of the contractor as it progresses.
- Upon completion of the work, evaluating whether or not the contractor has met the requirements of the construction contract and whether or not the owner has received value for the payments made to, or to be made to, the contractor.

Engineer's Duties and Authority

The employer appoints the engineer who shall carry out the duties assigned to him in the contract. The engineer administers the contract and has no authority to amend it.

The engineer exercises his authority as specified in or necessarily to be implied from the contract.

He is generally required to obtain the employer's approval before exercising the following:

1. The instruction of any variation or assessment of any application for an extension of time for completion or for any cost associated with a variation.
2. Any amendment to the programme.[25]
3. The issue of any taking over certificate or performance certificate.
4. The issue of an instruction to suspend the works or terminate the contract under the contract.

Other performances of the engineer in a construction contract include:

1. Whenever carrying out duties or exercising authority, specified in or implied by the contract, the engineer shall be deemed to act for and on behalf of the employer.
2. The engineer has no authority to relieve either party of any duties, obligations or responsibilities under the contract.
3. Any approval, check, certificate, consent, examination, inspection, instruction, notice, proposal, request, test or similar act by the engineer shall not relieve the contractor of any responsibility he has under the contract, including responsibility for errors, omissions, discrepancies and non-compliance.

[25] Schedule—being baseline, revised or updated schedules.

Ethics for Engineers

Engineering is an important and learned profession. As members of this profession, engineers are expected to exhibit the highest standards of honesty and integrity. Engineering has a direct and vital impact on the quality of life for all people. Accordingly, the services provided by engineers require honesty, impartiality, fairness and equity, and must be dedicated to the protection of public health, safety and welfare.

Engineers must perform under a standard of professional behaviour that requires adherence to the highest principles of ethical conduct.

Engineers, in the fulfilment of their professional duties, shall:

1. Hold paramount the safety, health, and welfare of the public.
2. Perform services only in areas of their competence.
3. Issue public statements only in an objective and truthful manner.
4. Act on behalf of each employer as faithful agent.
5. Avoid deceptive acts.
6. Conduct themselves honourably, responsibly, ethically, and lawfully so as to enhance the honour, reputation, and usefulness of the profession.

Engineers as well shall:

1. Be guided in all their relations by the highest standards of honesty and integrity.
2. At all times strive to serve the public interest.
3. Avoid all conduct or practice that deceives the public.
4. Not disclose, without consent, confidential information concerning the business affairs or technical processes of any present or former employer or employer, or public body, on which they serve.
5. Not be influenced in their professional duties by conflicting interests.
6. Not attempt to obtain employment or advancement or professional engagements by untruthfully criticising other engineers, or by other improper or questionable methods.
7. Not attempt to injure, maliciously or falsely, directly or indirectly, the professional reputation, prospects, practice or employment of other engineers. Engineers who believe others are guilty of unethical or illegal practice shall present such information to the proper authority for action.
8. Accept personal responsibility for their professional activities—provided, however, that engineers may seek indemnification for services arising out of their practice for other than gross negligence, where the engineer's interests cannot otherwise be protected.
9. Give credit for engineering work to those to whom credit is due, and recognise the proprietary interests of others.

Entire Agreement

An entire agreement clause aims to ensure that all the terms and conditions governing the rights and obligations of the parties are set out in a single contractual document, superseding all prior negotiations and agreements. The goal of such a clause is to prevent contracting parties from relying upon statements or representations made by them during negotiations for the purposes of claiming that they had agreed to something different than what is stated in the contract at the time of a dispute.

Entire agreement clauses are intended to do so by nullifying legal causes of action and limiting the contract to the expressed terms of the contract.

An entire agreement clause provides that the agreement is limited to the materials referred to in the contract, thus excluding prior contracts, informal or formal arrangements already negotiated, and any previous memorandum of understandings (McKendrick 2020).

Entire agreement clauses have received close judicial scrutiny over many, many years. The long line of cases on entire agreement clauses show that entire agreement clauses are often related to the background of the contract, the terms of the contract as a whole, and the precise terms of the entire agreement clause, within the context of the rest of the contract.[26]

In a typical construction contract, the contract shall constitute the entire agreement between the parties in relation to the works and shall supersede and extinguish any previous agreements, arrangements, understandings and representations relating thereto that are not set out in the contract.

In a standard entire agreement clause the contractor warrants to the employer that, in entering into the contract, it does not rely on any statement, representation, assurance or warranty of any person (whether a party to the contract or not and whether oral or in writing) other than as expressly set out in the contract.

Fitness-for-Purpose

Essentially, a fitness-for-purpose obligation means that a contractor warrants the thing he has designed or supplied will be fit for its intended purpose. This is a very onerous requirement, because it potentially imposes a strict liability on the contractor or designer to ensure a particular outcome, irrespective of any outside influences which could make it difficult to comply.[27]

Under many contract conditions, the contractor will usually find itself subject to a fitness-for-purpose obligation in respect of all works and design issued to him.

---

[26] *AXA Sun Life Services Plc v Campbell Martin Ltd & Ors [2011] EWCA Civ 133.*

[27] *Greaves v Baynham Meikle [1975] 1 WLR 1095 at 1098, CA.*

Under law, the fitness-for-purpose duty is stricter than the ordinary responsibility of an architect or other consultant carrying out design where the implied obligation is one of reasonable competence to 'exercise due care, skill and diligence'.[28]

The contractor usually is not liable for defects in the works due to his design insofar as he proves that he used reasonable skill and care to ensure that his design complied with the works information, unless it failed to carry out that design using the skill and care normally used by professionals designing works similar to the works.

A common issue of dispute between parties who have entered into a design and build contract is that quite often, the design portion and the build portion of the contract give rise to different obligations. Contractors are often unaware of this and/or incorrectly presume these obligations have the same consequence and are freely interchangeable.

As a contractor, when considering a contract, he or she should be clear on where liability for design and liability for workmanship begins and ends. A fit for purpose obligation effectively warrants a particular performance for the life of the project being built. This is clearly much more onerous than a simple reasonable care and skill obligation and should therefore be avoided if possible.

Force Majeure

Force majeure means, in relation to any party, any circumstances beyond its reasonable control affecting the performance by the party of its obligations under an agreement including adverse weather conditions, serious fire, storm, flood, lightning, earthquake, explosion, acts of a public enemy, terrorism, war, military operations, insurrection, sabotage, civil disorder, epidemic/pandemic, embargoes, labour disputes of a person other than said party, or Acts of God.

The term *force majeure* has been defined as

> an event or effect that can be neither anticipated nor controlled. It is a contractual provision allocating the risk of loss if performance becomes impossible or impracticable, especially as a result of an event that the parties could not have anticipated or controlled. (Garner 2019).

A *force majeure* clause in a contract would typically:

- Render impossible the affected party's performance of its obligations under the contract.
- Be beyond the control of the affected party and not due to its acts or omissions.
- Not have been prevented, avoided or overcome by the affected party through the exercise of due diligence.

Force majeure shall not include:

[28] *H W Nevill (Sunblest) v William Press [1981] 20 BLR 78.*

1. Any shortage or late delivery of plant and/or materials or consumables that the contractor or subcontractor is obliged to supply under the contract or subcontracts and/or in connection with the works.
2. Any shortage of staff and labour.
3. Any act of a governmental authority in relation to permits.
4. Changes in market conditions.
5. An inability to secure financing (and the effects thereof) for whatever reason or due to an insolvency event occurring, in relation to the contractor, any subcontractor or supplier of the contractor, any subcontractor or supplier of any tier, or a guarantor or financial institution providing any performance securities and/or bonds.

The usual remedy (which again can be varied by contract) if a force majeure event is proven is that performance of the affected obligation(s) is suspended until such time as the *force majeure* event ceases to impact performance.

A party who has the benefit of a contract that contains *force majeure* provisions is unlikely to seek to rely on the common law doctrines of impossibility or frustration. *Force majeure* provisions are likely to offer more certainty and the court will seek to apply the terms of the contract.

In the event of *force majeure*, the contractor shall continue to use its best endeavours to complete the execution of the works. Further, the contractor shall at all times use its best endeavours to minimise any cost and delay in the performance of the contract as a result of *force majeure*.

Frustration

The courts have typically regarded the items on the following (non-exhaustive list) to be frustrating events:

- Destruction of the subject matter of the contract.
- Supervening illegality (i.e. when a law subsequent to the contract is passed which renders the fundamental principal of the contract illegal).
- Incapacity or death of one of the parties.
- Serious delay that affects the intended purpose of the contract.

A basic test for frustration in construction contract was set out by in *Lord Radcliffe in Davis Contractors v Fareham UDC*,[29] resulting in the three basic points:

1. A frustrating event is not caused by the default of either party.
2. The contract becomes impossible to fulfil as it has become something entirely different from the original agreement between the parties.
3. There is no provision in the contract to cover the eventuality that is a frustrating event.

[29] *[1956] AC 696.*

Frustration or impossibility operates to 'kill the contract' and if a contract is found to be frustrated or impossible then the parties are discharged from further performance of their obligations.

For this reason, courts are reluctant to apply the doctrines and have applied them in only limited circumstances.

For a party to be able to rely on the doctrine of frustration, it will need to establish that performance of the contract is genuinely impossible, rather than just more difficult or expensive. It is not enough that performance would merely inflict extreme, even ruinous, hardship on the performing party. If there is a way of performing the contract in something approaching the manner originally contemplated by the parties, that must be done, irrespective of the burden. The doctrine of frustration is notoriously difficult to successfully establish, and the consequences of its application may be harsh and unpredictable.

Good Faith

One of the potential difficulties with international projects is that the contracts entered into are governed by laws which may be unfamiliar to one or other of the contracting parties. Therefore, it is important that parties to a contract do not make assumptions either as to what particular clauses mean or as to which legal principles they can imply into that contract, as sometimes particular jurisdictions take an entirely different approach to the one that you might have been expecting.

The principles of good faith imply that in exercising his rights and performing his duties, each party must act in accordance with good faith and fair dealing. The intention is that this rule runs through the entire contract, from negotiation to final ultimate completion. However, that does not actually assist in establishing what good faith actually is. And this uncertainty is one of the main reasons why the courts are reluctant to deal with the concept.[30]

Generally, to act in good faith, each party in a construction contract is under an obligation to:

1. Not misrepresent material facts.
2. Disclose material facts even if no question has been raised about them.

Law jurisdictions recognise a duty of good faith, if the contract contains an express term to that effect. Traditionally, however, common law has not recognised an obligation of good faith as implicit in all contracts, or even in all commercial contracts. In *Interfoto Picture Library v Stiletto Visual Programmes*[31]; Bingham LJ famously stated:

> In many civil law systems, and perhaps in most legal systems outside the common law world, the law of obligations recognises and enforces an overriding principle that in making and

[30] *Yam Seng Pte Ltd v International Trade Corporation Ltd [2013] EWHC 111.*

[31] *[1987] EWCA Civ 6.*

> carrying out contracts parties should act in good faith. This does not simply mean that they should not deceive each other, a principle which any legal system must recognise; its effect is perhaps most aptly conveyed by such metaphorical colloquialisms as 'playing fair', 'coming clean' or 'putting one's cards face upwards on the table'.

Some US courts have accepted that all contracts have an implied good faith obligation. Other common law jurisdictions are moving in the same direction.

Indemnity

The term 'indemnity' literally means 'security against loss'. In a contract one party (the indemnifier) promises to compensate the other party (the indemnified) against the loss suffered by the other.

Indemnity is a contractual agreement between two parties, which outlines a form of insurance compensation for any damages and losses. In an indemnity agreement, one party will agree to offer financial compensation for any potential losses or damages caused by another party, and to take on legal liability for whatever damages were incurred.

The contractor is usually responsible for the care of the works, from the commencement date until takeover, but he is also to a certain extent responsible for things that arise out of or as a consequence of his execution of the works or remedying of any defects in terms of:

1. Death or injury to any person.
2. Loss or damage to any property (other than the works) as a consequence of the performance of the contractor's duties.

However, while the contractor must indemnify the employer for losses or claims for bodily injury, disease or death of any person, regardless of whether or not the contractor was negligent (unless positively caused by the employer or his agents), the contractor is only liable to indemnify the employer for damages where the contractor has been negligent or committed a breach of contract.

An indemnity clause in a construction contract between the employer and the contractor may protect an employer from liability to the extent that someone is injured on the site during a construction project. For a contractor, however, it is essential that the construction contracts are drafted in a manner to include an indemnity clause that protects the contractor's interests, in case an employer or subcontractor asserts a claim against him.

In a typical clause; the contractor shall defend, indemnify and hold the indemnified parties harmless against any and all losses, liabilities, damages, fines, costs, expenses (including legal fees), demands, claims, actions or proceedings which the indemnified parties may suffer or incur in respect of:

1. Bodily injury, sickness, disease or death, of any person.
2. Damage to or loss of any real or personal property.

3. Any third parties including a lender in respect of a loan or other financing facility.
4. An intellectual property infringement claim pursuant contract.

Liquidated Damages and Penalty

Liquidated Damages

In construction contracts, liquidated damages are most often used in respect of the contractor not completing the works on time, and because of that breach, the employer seeking to recover liquidated damages on the basis of a certain fixed financial amount. A liquidated damages clause helps to eliminate uncertainty because the parties know in advance the financial liability that will arise if the contractor fails to complete the works on time.

This acts as a protection for the contractor against un-liquidated general damages claims and enables the contractor to fix the price for risk.

For an employer, a liquidated damages clause avoids the difficulty, time and expense involved in proving and assessing the actual loss that a party will suffer in the event of a breach.

The courts have found that a liquidated damages clause is only enforceable if it represents a genuine pre-estimate of the losses[32] that are likely to be incurred as a result of breach of contract.[33]

There are a number of grounds for which a liquidated damages clause may be held to be unenforceable, including:

1. Ambiguity; as a general rule, liquidated damages clauses must be constructed strictly *contra proferentem*. Therefore, where the clause is ambiguous and all other methods of construction have failed to resolve the ambiguity, the court may construe the words against the party seeking to rely on the clause.
2. Its penal nature; i.e. it is regarded as penalty in view of the courts.
3. The failure to provide for an extension of time in cases where the employer prevents or delays completion. This falls under Prevention Principle.
4. The failure to comply with contractual procedures[34] (for example which related to Sectional Completions).

[32] Usually set at no more than 10%.

[33] *Regional Construction Sdn. BHd. v Chung Syn Kheng Electrical Co. Bhd. [1987] 2 MLJ 763.*

[34] *Bramall & Ogden v Sheffield City Council (1985) 1 Con LR 30.*

Penalty v Liquidated Damages

The distinction between liquidated damages and a penalty is a matter of construction for the court, and it is for the party from whom the damages are claimed to show that the clause is in fact a penalty.

A liquidated damages clause that is found to be penal is generally invalid. It is a well-accepted principle that liquidated damages provisions must, in both their amount and operation, constitute a genuine pre-estimate of the loss that the employer is likely to suffer as a result of the breach. It should be noted that the pre-estimate does not need to bear any relationship to the actual damage suffered as a consequence of the breach.

If a liquidated damages clause is found to be penal in nature, the clause will be invalid and unenforceable and the court will only enforce the sum identified where it represents a proper assessment of the loss.

The use of the words penalty or liquidated damages in the contract is not conclusive or determinative of whether the liquidated damages clause is in fact one or the other.

The provisions to consider when addressing penalties as to liquidated damages in construction contracts can be described by the following criteria:

1. The essence of a penalty is a payment of money stipulated against the contractor whereas the essence of liquidated damages is a genuine estimate of damages to the employer. The key is to attempt to estimate what actual loss was caused by the breach and whether it is fair and reasonable.
2. The question whether a sum stipulated is a penalty or is liquidated damages is a question of construction to be decided upon the terms and inherent circumstances of each particular contract. To assist this task of construction of such provisions various tests have been devised[35]:

(a) It will be held to be a penalty if the stipulated sum is extravagant and unconscionable in amount in comparison with the greatest loss that could conceivably be proved to have followed from the breach.

(b) It will be held to be a penalty if the breach consists only in paying a sum of money, and the sum stipulated is a sum greater than the sum that ought to have been paid (i.e. greater than the actual losses of the innocent party).

(c) There is a presumption that it is penalty when a single lump sum is made payable by way of compensation, on the occurrence of one or more or all of several events, some of which may occasion serious and others but trifling damage.

(d) The time for assessment or construction of the provision as either a genuine pre-estimate of damages or a penalty is as at the time of the contract. In other words, if it is a genuine pre-estimate of damages or losses of the employer at the time the contract is entered, then it is likely to be valid and enforceable.

---

[35]*Dunlop Pneumatic Tyre Co v New Garage & Motor Co Ltd ([1915] Ac 79).*

Negligence

Tort and Causation

Damages for breach of contract are not the only means by which general principles of law allow recovery of monetary compensation. There are a variety of other general principles that can also provide monetary compensation, depending on the circumstances.

In the construction industry, one of the most commonly relied upon principles is the law of negligence. The tort of negligence is not concerned with a breach of a contract, but with wrongful acts. In fact, there need not be a formal contract at all, as long as it can be shown that one party owed a duty of care to another and that the duty of care has been breached. The term negligence is found in the context of breach of contract, for example, where an architect is alleged to have carried out negligent design or supervision.[36]

Negligence is also a failure to take reasonable care for the safety or well-being of others. Negligent actions are not an exercise in perfection but rather address issues of reasonableness or, put simply, what a reasonable party might have done or not done in the circumstances of a particular case. The law of negligence entitles a party to receive compensation, for loss or damage, as a result.

In general terms, negligence can be established if:

- the defendant owed them a duty to take reasonable care; and
- the defendant breached that duty; and
- the defendant's breach of duty caused the injury or damage suffered by the plaintiff; and
- the injury or damage was not too remote a consequence of the breach of duty.[37]

Duty of Care and Negligent Statement

Duty of care, stated simply, means that one must take reasonable steps to ensure one's actions do not knowingly cause harm to another individual. In such cases, the courts look to the nature of the relationship between the parties and whether the incident resulting in harm was reasonably foreseeable.[38]

It is also imperative that there is proximity or causal connection between one person's conduct and the other person's injury. If the actions of a person are not made with watchfulness, attention, caution and prudence then their actions are considered

[36] *Blyth v Birmingham Waterworks* Company (1856) 11 Ex Ch 781.

[37] *JEB Fasteners Ltd v Marks, Bloom & Co (1982); Morgan Crucible v Hill Samuel Bank Ltd (1991); James McNaghten Paper Group Ltd v Hicks Anderson & Co (1991).*

[38] *Donoghue v Stevenson (1932) AC 562.*

negligent; consequently, the resulting damages may be claimed as negligence in a lawsuit (Merkin and Saintier 2019).

Historically, the accuracy of the statement would be warranted as a term of the contract. If the statement was incorporated in the contract there could be liability for breach of contract. Financial relief was only available if the statement was incorporated in the contract or had been made dishonestly. Such a term would be classified as a 'promissory representation' or simply a mis-description.

The law was very reluctant to allow a party to claim damages for losses suffered as a result of a statement being untrue, unless the maker of the statement had made it as part of a contract or had made the statement fraudulently. Compensatory remedies were only available in respect of misrepresentations incorporated in a contract, in which case damages would be recoverable for breach of warranty, or in respect of losses suffered as a result of a fraudulent statement, in which case damages were recoverable in the tort of deceit.[39]

Hedley Byrne & Co. Ltd. v Heller & Partners Ltd[40] was a watershed in regard to the doctrines of reasonableness, duty of care and negligent misstatement. A feature of this case is that there was an approach, made to the defendant bank by or on behalf of the plaintiffs, inviting the bank to provide a service of advice and information directly to them. Lord Pearce held that there could be an implied duty of care imposed on professionals acting on behalf of employers in giving statements and advice. In this regard, he stated the following:

> A duty of care created by special relationships which, though not fiduciary, gives rise to an assumption that care as well as honesty is demanded.

Furthermore, as to the issue of duty of care imposed on a professional providing a statement, Lord Reid said:

> A reasonable man, knowing that he was being trusted or that his skill and judgement were being relied on, would, I think, have three courses open to him. He could keep silent or decline to give the information or advice sought: or he could give an answer with a clear qualification that he accepted no responsibility for it or that it was given without that reflection or inquiry which a careful answer would require: or he could simply answer without any such qualification. If he chooses to adopt the last course he must, I think, be held to have accepted some responsibility for his answer being given carefully, or to have accepted a relationship with the inquirer which requires him to exercise such care as the circumstances require.

In *Hedley Byrne & Co. Ltd. v Heller & Partners Ltd*, Lord Morris, as to when men of special skills have a duty of care implied in their statements, further stated:

> My lords, I consider that it follows and that it should now be regarded as settled that if someone possessed of a special skill undertakes, quite irrespective of contract, to apply that skill for the assistance of another person who relies on such skill, a duty of care will arise. The fact that the service is to be given by means of or by the instrumentality of words can make no difference. Furthermore if, in a sphere in which a person is so placed that others could reasonably rely on his judgement or his skill or on his ability to make careful inquiry,

[39] *Caparo Industries PLC v Dickman [1990] UKHL 2.*

[40] *[1964] AC 465.*

> a person takes it on himself to give information or advice to, or allows his information or advice to be passed on to, another person who, as he knows or should know, will place reliance on it, then a duty of care will arise.

Prevention Principle

The essence of the prevention principle is that the employer cannot hold the contractor to a specified completion date if the employer has, by an act or omission, prevented the contractor from completing by that date. Instead, time becomes at large and the obligation to complete by the specified date is replaced by an implied obligation to complete within a reasonable time.

The leading case on prevention principle is *Holme v Guppy*[41] where the employer delayed granting the contractor access to the site. Parke B held:

> if the party be prevented, by the refusal of the other contracting party, from completing the contract within the time limited, he is not liable in law for the default … The plaintiffs were excused from performing the agreement contained in the original contract … and consequently they are not to forfeit anything for the delay.

It was also stated by Lord Denning,[42] one of the twentieth century's most influential jurists, that

> it is well settled that in building contracts—and in other contracts too—when there is a stipulation for work to be done in a limited time, if one party by his conduct—it may be quite legitimate conduct, such as ordering extra work—renders it impossible or impracticable for the other party to do his work within the stipulated time, then the one whose conduct caused the trouble can no longer insist upon the strict adherence to the time stated. He cannot claim any penalties or liquidated damages for non-completion in that time.

This notion has come to be known as the prevention principle. In the field of construction law, one consequence of the prevention principle is that the employer cannot hold the contractor to a specified completion date, if the employer has by act or omission prevented the contractor from completing by that date. Instead, time is set at large; the contractor is obliged to complete within a 'reasonable time'; and the employer loses its right to claim liquidated damages for delay.

Instead, to claim any damages, the employer will need to prove both that:

- The contractor took longer than a 'reasonable time'.
- The employer suffered particular losses as a result.

The leading authority on this point is the decision in *Peak Construction (Liverpool) Ltd v McKinney Foundations Ltd*.[43] In that case, Lord Justice Salmon confirmed that a liquidated damages clause could not apply and further commented:

> I cannot see how, in the ordinary course, the employer can insist upon compliance with a condition if it is partly its own fault that it cannot be fulfilled.

[41] *(1838) 3 M & W 387.*

[42] *Trollope & Colls Ltd v North West Metropolitan Regional Hospital [1973] 1 W.L.R. 601.*

[43] *(1970) 1 BLR 111.*

Privity

Privity of contract is a fundamental principle in contract law, meaning that only the parties to a contract can enforce its terms. A third party cannot, save in exceptional cases, enforce a contract to which it is not a party—it had no 'rights' in respect of that contract.

The doctrine of privity of contract was developed by the common law because common law focuses more on the issue of who is entitled to sue for damages, rather than who derives rights under the contract.[44]

As to the principles of privity, the common law reasoned that:

1. There is the principle that consideration must move from the promisee. See *Tweddle v Atkinson*.[45]
2. Only a promisee may enforce the promise, meaning that if the third party is not a promisee he is not privy to the contract. See *Dunlop Tyre Co v Selfridge*.[46]
3. A contract between two parties may be accompanied by a collateral contract between one of them and a third person relating to the same subject-matter. See *Shanklin Pier v Detel Products*.[47]

The doctrine of privity of contract in the context of construction contracts applies in subcontracting and professional services contracts.

By operation of the doctrine a subcontractor, even a nominated subcontractor, contracts with the main contractor and has no privity of contract with the employer. Therefore, the subcontractor has no recourse whatsoever against the employer as there is no contractual nexus between them. Likewise, the contractor has no recourse against an agent appointed by the employer in terms of a contract, as he is usually not party to the professional services contract between the agent and the employer.

Programme

The initial contract programme,[48] called the baseline programme, shall be submitted by the contractor for approval within seven (7) days[49] of the award of the project work to the engineer, and shall clearly identify the early and late start and finish dates of each activity and the activities' interrelationship with other activities.

---

[44]Merkin, R., Saintier,S.; Poole's *Casebook on Contract Law,* Oxford University Press, 14th edn., 2019.

[45]*(1861) 1 B&S 393*. The fathers of a husband and wife agreed in writing that both should pay money to the husband, adding that the husband should have the power to sue them for the respective sums. The husband's claim against his wife's father's estate was dismissed, the court justifying the decision largely because no consideration moved from the husband.

[46]*[1915] UKHL 1.*

[47]*[1951] 2 KB 854.*

[48]Also called 'schedule'.

[49]Or as dictated in the contract, but usually no more than 28 days.

In the baseline programme, the critical path shall be identified separately and the float time shown. Once approved by the engineer, the contractor shall be responsible for maintaining, monitoring and submit proposed revisions to the contract baseline programme due to changes made during the performance of the work as approved by the employer.

All initial information presented will be updated and kept up-to-date. There shall be no change in the baseline programme without prior written approval from the engineer or the employer.

Costs associated with keeping the baseline programme up-to-date are the contractor's responsibility.

The engineer is required to review the programme and confirm if it does not comply with the contract.

If the engineer does not do this within the specified date in the contract, then the programme is deemed to be approved. There is also a positive obligation on the contractor to update the programme whenever it ceases to reflect actual progress.

Provisisonal Sum

A provisional sum is an allowance included in a fixed price construction contract for an item of work that cannot be priced by the contractor at the time of entering the contract.

Moreover, a provisional sum is a sum provided either for defined works or for undefined work:

1. *Defined Provisional Sum* This is defined as when work is not completely designed at the time of issuing the tender document, but specific information can be provided.
2. *Undefined Provisional Sum* Undefined provisional sum, on the other hand, is related to work for which such information cannot be given.

Each provisional sum shall only be used, in whole or in part, in accordance with the engineer's instructions, and the contract price shall be adjusted accordingly.

The total sum paid to the contractor shall include only such amounts, for the work, supplies or services to which the provisional sum relates, as the engineer has instructed. For each provisional sum, the engineer may instruct:

1. Work to be executed (including plant, materials or services to be supplied) by the contractor.
2. Plant, materials or services to be purchased by the contractor, for which there shall be added to the contract price less the original provisional sum.
3. The actual amounts paid (or due to be paid) by the contractor.
4. A sum for overhead charges and profit, calculated as a percentage of these actual amounts by applying the relevant percentage rate (if any) stated in the contract.

The contractor shall, when required by the engineer, promptly produce quotations, invoices, vouchers and accounts or receipts in substantiation.

The contractor shall not usually be entitled to an extension of the time for completion in respect of any provisional sum. The contract price will be adjusted to take account of the expenditure of the provisional sum but no adjustment shall be made to the percentage for contractors' overhead and profit, or to the amounts for general attendance. The value of the provisional sum will be taken to include these items.

Risks

Provisions for the allocation of risk among parties to a contract can appear in numerous areas in addition to the total construction price. Typically, these provisions assign responsibility for covering the costs of possible or unforeseen occurrences. A partial list of responsibilities with concomitant risk that can be assigned to different parties would include:

- Force majeure (i.e. this provision absolves an owner or a contractor for payment for costs due to 'Acts of God' and other external events such as war or labour strikes).
- Indemnification (i.e. this provision absolves the indemnified party from any payment for losses and damages incurred by a third party such as adjacent property owners).
- Liens (i.e. assurances that third-party claims are settled such as 'mechanics' liens' for workers' wages).
- Labour laws (i.e. payments for any violation of labour laws and regulations on the job site).
- Differing site conditions (i.e. responsibility for extra costs due to unexpected site conditions).
- Delays and extensions of time.
- Liquidated damages (i.e. payments for any facility defects with payment amounts agreed to in advance).
- Consequential damages (i.e. payments for actual damage costs assessed upon impact of facility defects).
- Occupational safety and health of workers.
- Permits, licenses, laws and regulations.
- Equal employment opportunity regulations.
- Termination for default by the contractor.
- Suspension of work.
- Warranties and guarantees.

Subcontractors

A subcontractor is a person or entity who has a direct contract with the contractor to perform a portion of the work at the site.

Unless otherwise stated in the contract documents or the bidding requirements, the contractor, as soon as practicable after award of the contract, shall furnish in writing to the employer through the engineer the names of subcontractors for each principal portion of the work, including parts of the scope of works such as (1) heating, ventilating and air conditioning; (2) plumbing; (3) electrical and (4) general.

The engineer shall promptly reply to the contractor in writing stating whether or not the employer or the engineer, after due investigation, has reasonable objection to any such proposed subcontractor.

In a typical subcontract agreement, a subcontractor has a contract with the contractor for the services provided.

The contractor shall not subcontract the whole or any part of the works without the prior written consent of the engineer, and this consent may be withheld by the engineer in its absolute discretion.

If the contractor wishes to enter into a subcontract or supply contract for any part of the works, the contractor shall:

1. provide the engineer with such information in relation to each proposed subcontractor or supplier and the relevant work package as the engineer may reasonably require;
2. obtain the engineer's approval of the subcontractor or supplier, the terms and conditions of any proposed subcontract or supply contract (including prices set out therein) and, if requested by the engineer, all associated tender documentation; and
3. provide to the employer, upon the employer's request, the tenders submitted by all tenderers for any subcontract, quotations from prospective suppliers for any supply contract, any amendments thereto and details of any negotiations held by the contractor with any tenderers and prospective suppliers.

At the same time as entering into a subcontract, including for any of the parts of the works, the contractor shall procure that the relevant subcontractor enters into a collateral warranty in favour of the employer in respect of the part of the works to be undertaken by the subcontractor.

The contractor shall give the engineer a notice of the intended commencement of each subcontractor's work on the site.

The contractor is fully responsible for the acts, omissions or defaults of any and all subcontractors and their agents or employees, as if they were acts, omissions or defaults of the contractor.

Sufficiency of the Contract Price

Because of the unique nature of constructed facilities, it is almost imperative to have a separate price for each facility. The construction contract price includes the direct project cost, including field supervision expenses plus the mark-up imposed by contractors for general overhead expenses and profit (Mason 2016).

In order for the contractor to practise utmost due diligence, and be comfortable with the sufficiency of his price, he must ensure that he does the following:

1. visit the site and allow for any risks associated with site conditions;
2. study tender drawings sufficiently;
3. review all items in the bill of quantities and ensure that the quantities match the quantities in the drawings;
4. study the specifications and make sure that there are no discrepancies between the specifications and the drawings;
5. study his overheads, including his site and office indirect costs;
6. calculate carefully his labour force's man-hours;
7. obtain quotations from subcontractors and from the supplier he will use; and
8. allow for contingencies.

The contractor must also make sure that with respect to the works of the contract he has:

1. examined all information relevant to the risks, contingencies and other circumstances having an effect on the contract price and the time for completion;
2. based the contract price on his own determinations and assessment of the risks involved in carrying out the works;
3. satisfied himself as to the correctness and sufficiency of the contract price for the works under the contract, and ascertained that the contract price covers the cost of complying with all of the contractor's obligations under this contract;
4. examined all applicable laws and permits relevant to the works and all orders, regulations, rules, directions and practices in relation to access to the work;
5. highlighted all security issues, which may restrict or inhibit the execution of any part of the works; and
6. satisfied himself of his capacity to execute the works in accordance with the terms and conditions of the contract without breaching any such terms and conditions.

It is a precondition to all contracts that the contractor must agree that with respect to the works agreed, the contract price covers all of the contractor's obligations under this contract (including those under provisional sums, if any), as well as all things necessary for the proper execution and completion of the works and the remedying of any defects therein.

Suspension of the Works

Most construction contracts will contain provisions setting out the parties' rights and obligations when the principal seeks to suspend the works. The situation is not always clear cut, however.

The employer, by a written order, may direct the contractor to suspend the work, or any portion thereof, in any of the following cases until the cause for such order has been eliminated:

1. Unsuitable weather or other conditions considered unfavourable for the execution of the work.
2. Failure of the contractor to correct conditions that constitute a danger to his workers or the general public, or to correct defective work.
3. Failure of the contractor to carry out valid orders issued by the employer or to comply with any provision of the contract, or his persistent failure to carry out the works in accordance with the contract.
4. The necessity for adjusting the drawings to suit site conditions found during construction, or in case of a change in drawings and specifications.
5. Failure of the employer to supply the employer's materials on time, where such failure is due to causes beyond the reasonable control of the employer.
6. Delay by the employer in obtaining a right of way, where such obligation is assumed by the employer under the contract, and the delay is not due to the fault or negligence by the employer.
7. *Force majeure* or fortuitous event.

The contractor shall immediately comply with such an order to suspend the work or any part thereof for such a period or periods and in such manner as the employer may direct, and during said suspension shall properly protect and secure the works.

In many contracts, the engineer may at any time instruct the contractor to suspend progress of part or all of the works through a suspension notice. During such a suspension, the contractor shall protect, store and secure said part or all of the works against any deterioration, loss or damage, and shall comply with any reasonable directions from the engineer in relation to protecting, storing and securing said part or all of the works.

Upon receipt of a suspension notice, the contractor shall within ten (10) days,[50] or as instructed by the engineer or the employer, after receipt of the suspension notice submit to the engineer, for its approval, the contractor's plan for the protection, storage and security of the works, or part of the works, suspension plan.

[50] Or as stated in the contract.

Taking Over by Employer and Completion of Milestone Works

During completion of the works, the handover of the site (at the stage of substantial completion or practical completion) to the employer takes place once the engineer has confirmed that the works defined in the contract are complete. The handover should be planned well in advance, and any special requirements must be included in the handover documents.

Substantial completion of the whole of the works and/or sections of it shall only be considered to have been reached, and the whole of the works and/or sections of it shall only be considered by the engineer ready for taking over, when the following conditions are satisfied:

1. The works or their relevant section have been completed in accordance with the requirements of the contract, subject only to minor outstanding items of work and/or rectification of minor blemishes and/or defects.
2. The existence of the said outstanding items of work and/or defects, individually and collectively, will not affect the use of the works or section for the purpose intended, and these issues can be completed and/or rectified very quickly and can be conveniently attended to by the contractor, after he has taken over, without causing any significant inconvenience or nuisance.
3. The contractor has provided the employer with the as-built documents, the operation and maintenance manuals and training that the contractor is obliged to provide to the employer pursuant under the contract.
4. All applicable laws regarding the use of the works or the relevant section have been complied with, and all permits that are required have been issued certifying that the works or relevant section(s) can be occupied and used for their intended purpose.
5. All collateral warranties requested by the employer have been executed and provided in favour of the parties identified by the employer.
6. The works or the relevant section of the works have passed the respective tests prior to completion, and duly certified reports of the tests and (unless the engineer instructs otherwise) any certificates, permits and reports in respect thereof that are required to be obtained from any relevant governmental authority or utility company have been forwarded to the engineer, and the engineer has endorsed the contractor's test certificate or issued a certificate to the contractor to that effect.
7. All sums due and payable by the contractor to the employer and its subcontractors and suppliers on the date on which the engineer issues the taking over certificate(s), have been paid in full.
8. Any other stipulations identified in the contract as being prerequisites to substantial completion and taking over have been satisfied.

After taking over, the following will be the responsibility of the contractor:

1. Completion of outstanding works and remedying defects, which includes the completion of any work that is outstanding or that requires rectification on the date stated in a taking over certificate, and carrying out of any outstanding tests prior to completion, either by the dates or within the time periods set out in the specification or within such time as is instructed by the engineer (at his absolute discretion) at the risk and cost of the contractor.
2. Execution of all works required to remedy any defects, deficiencies and/or damage discovered after the issuing of a taking over certificate, as may be notified by the engineer either by the dates or within the time periods set out in the specification or within such time as is instructed by the engineer (at his absolute discretion) and in any event on or before the expiry date of the defects notification period.
3. All remedial work shall be executed at the risk and cost of the contractor, unless such work is attributable to any breach of the contract by the employer, in which case the engineer may give notice to the contractor.

If the contractor fails to complete the outstanding work or remedy the defect, deficiency or damage by the notified date referred, the employer may (at its discretion):

1. carry out the work itself or have it carried out by others, in a reasonable manner and at the contractor's cost, and the contractor shall pay to the employer the costs incurred by the employer in remedying the defect, deficiency or damage; or
2. require the engineer to agree or determine a deduction from amounts otherwise due to the contractor, or require the engineer to determine such a deduction.

Termination

Termination by Employer

Termination clauses give parties the right to terminate the contract in certain circumstances as determined by the contract signed between the parties. Termination clauses most commonly deal with breaches of specified contractual obligations.

Under a common contract; the employer may by notice terminate the contract immediately if:

1. An insolvency event occurs in relation to the contractor, or the employer (acting reasonably) is of the opinion that an insolvency event is likely to occur in relation to the contractor;
2. The contractor, a subcontractor, an affiliate of the contractor or any subcontractor, or any of the contractor's personnel, commits a prohibited act; or

i. Fails to comply with a maintenance of bonds clause.
ii. Fails to comply with a notice to correct clause.
iii. Abandons the works or otherwise plainly demonstrates its intention not to continue performance of its obligations under the contract.
iv. Without reasonable excuse, fails to proceed with the works or any section of them (as the case may be) in accordance with the commencement, delays and suspension clauses set out in the contract.
v. Subcontracts part of or the whole of the works without the consent of the employer required in accordance with a subcontracting and suppliers clause.
vi. Assigns the contract without prior consent from the employer.
vii. Fails to comply with clauses related to financial statements and a disclosure obligations clause.
viii. Fails to pay its subcontractors and/or suppliers.
ix. Makes any false disclosure.
x. Suspends the performance of the works or any part thereof other than as a consequence of *force majeure* or as necessary for the safety or security of the works.
xi. Persistently fails to execute the works in accordance with the requirements of the contract or persistently neglects to carry out and/or fulfil its obligations and/or responsibilities under the contract without due cause and the employer has given *fourteen (14) days'*[51] notice of its intention to terminate the contract unless the contractor is able to satisfy the employer that appropriate steps are being taken to ensure that there is no reoccurrence of such default.
xii. Is in breach of any material obligation that it has under the contract and the employer has given *fourteen (14) days'*[52] notice of its intention to terminate the contract unless the contractor remedies such breach to the satisfaction of the employer, and the contractor fails or is unable to remedy such breach to the satisfaction of the employer prior to the expiry of such notice.
xiii. Is unable to make, or fails to make, a material start to a section or a part of the works on-site within sixty (60) days[53] of the employer granting right of access to the relevant section or part of the site (other than as a result of *force majeure*).
xiv. After having commenced the works on-site, the contractor is unable or fails to proceed with the works with due expedition and without delay (other than as a result of *force majeure* or employer breach clauses).
xv. Fails to comply with applicable laws (except if instructed to do so by the engineer in writing) which exposes the employer to liability or adversely impacts upon the employer's ability to use, operate, maintain,

[51] Or another date inserted in the contract.
[52] Or another date inserted in the contract.
[53] Or another date inserted in the contract.

finance, lease or dispose of the completed works and the employer has given required notice of its intention to terminate the contract unless the contractor remedies such breach to the satisfaction of the employer, and the contractor fails or is unable to remedy such breach to the satisfaction of the employer prior to the expiry of such notice.

xvi. Fails to provide any one of the performance securities and/or bonds required under the contract when required to do so.

The employer's election to terminate the contract shall not prejudice any other rights of the employer, under the contract or otherwise.

Consequences of Termination for Contractor Default

If the contractor fails to carry out any obligation under the contract, the employer may by notice require the contractor to make good the failure and to remedy it within a specified time.

After a notice of termination has taken effect, and contractor has not remedied the elements of default under a notice the contractor shall promptly:

1. cease all further works, except for such works as may have been instructed by the employer for the protection of life or property or for the safety of the works;
2. deliver to the employer all contractor's documents;
3. deliver to the employer any required goods, including the hardware and software and all licences relating thereto;
4. remove all other goods and equipment from the site, except as necessary for safety, and leave the site; and
5. use its best efforts to comply immediately with procedures necessary for the protection of life or property or for the safety of the works.

As soon as practicable, after a notice of termination has taken effect, the engineer shall proceed accordingly to agree or determine the value of the works, goods and contractor's documents, and any other sums due to the contractor for works executed in accordance with the contract.

Termination for Employer's Convenience

In most contracts nowadays, there is a standard clause for the employer to terminate for convenience, i.e. for no reason, just an absolute power to terminate. by giving notice of such termination to the contractor.

After a notice of termination, under the related clause (Termination for Employer's Convenience),[54] has taken effect, the contractor shall promptly:

[54]Usually within a defined duration in the contract not exceeding 28 days.

1. cease all further work, except for such work as may have been instructed by the employer for the protection of life or property or for the safety of the works;
2. deliver to the employer all contractor's documents;
3. deliver to the employer all plants, materials and other works, for which the contractor has received payment;
4. remove all other goods and equipment from the site, except as necessary for safety, and leave the site; and
5. use its best efforts to comply immediately with any reasonable instructions included in the notice referred to in the related clause (Termination for Employer's Convenience) for the assignment or novation of any subcontract or consultant's appointment, and for the protection of life or property or for the safety of the works.

After a notice of termination for convenience has taken effect, the employer shall return the performance securities and bonds to the contractor and pay to the contractor:

1. the amounts payable for any work carried out by the contractor in accordance with the contract up to the date of termination, minus the amount of all payments previously paid to the contractor;
2. the cost of plant and materials ordered for the works that have been delivered to the contractor, or of which the contractor is liable to accept delivery. Such plant and materials shall become the property of the employer when paid for by the employer, and the contractor shall place them at the employer's disposal;
3. any other cost that in the circumstances was reasonably and unavoidably incurred by the contractor in the expectation of completing the works;
4. the cost of the removal of temporary works and the contractor's equipment from the site; and
5. the cost of demobilisation and repatriation of the contractor's staff and labour where such individuals were brought into the country solely for the purpose of, and remain employed wholly in connection with, the works, at the date of termination.

Termination by Contractor

The contractor shall be entitled to terminate the contract if:

- the employer fails to pay the contractor the amount due under any interim payment certificate or under the final payment certificate after the expiry of the time stated in certain clauses or schedules (i.e. the Contract Price, Payment Terms, and Payment Schedule clauses, as well as the Timing of Payments clause that states the period within which payment is to be made), and the contractor has served a notice on the employer following this protracted period advising the employer of its intention to terminate the contract unless payment is forthcoming;

- subject to the Cause for Suspension clause, whereby a prolonged suspension affects the whole of the works as described in clause (Prolonged Suspension); or
- an insolvency event occurs in relation to the employer.

The contractor's election to terminate the contract shall not prejudice any other rights of the contractor, under the contract or otherwise.

After a notice of termination under the Termination by Contractor clause has taken effect, the contractor shall promptly:

1. cease all further work, except such work as may have been instructed by the engineer for the protection of life or property or for the safety of the works;
2. deliver to the employer all the contractor's documents;
3. deliver to the employer all plants, materials and other works for which the contractor has received payment; and
4. remove all other goods from the site, except as necessary for safety, and leave the site.

Time at Large and Reasonable Time

The time for completion of works can become at large when the contractor has been hindered or prevented by the employer from completing the works in accordance with the original contract.[55] This results from a well-established principle of law that 'no person can take advantage of the non-fulfilment of a condition the performance of which has been hindered by himself'.

Time being at large does not mean that the contractor has no obligation to complete the work. He has to complete in a reasonable time and the employer is not entitled for liquidated damages.[56]

Time for completion of the works can be said to be, or made, at large in the following situations (Pickavance 2010):

1. No time for completion is fixed in the contract.
2. There is improper administration or misapplication of the extension of time provision in the contract.
3. There is a waiver of time requirements.
4. The extension of time provision does not confer power on the engineer to extend the time for completion of the works on the occurrence of an event or events that fall within the obligation of the employer.

The term 'time at large' is usually used in construction contracts in the situation where liquidated damages are an issue. If time is at large then it is argued liquidated damages cannot be applied, because there is no date fixed for their calculation. Late delivery cannot then be sanctioned by liquidated damages; instead the employer is

[55] Prevention Principle (supra).

[56] Even if the employer is not entitled to liquidated damages he can still recover general damages, if he can prove that he has suffered a loss as a result of the contractor delay.

faced with the problem of having to prove the damages he may want to claim, in addition to the larger issue of the works not being delivered on time.

Although it is thought of as a good negotiating bargain by contractors,[57] time being put at large also leaves them facing uncertainty as to the level of resources to implement and the monetary recovery for the extended works duration, as well as claims from the employer, which could end up more expensive than capped liquidated damages.

If the contractor incurs additional costs as a direct result of the employer's delay and/or contractor's delay, then the contractor should only recover monetary compensation if he is able to separate the additional costs caused by the employer's delay from those caused by the contractor's delay. Therefore, a contractor's delay should not reduce the amount of the extension of time due to the contractor as a result of the employer's delay, and analyses should be carried out for each event separately and strictly in the sequences in which they arise.

The following factors are potentially relevant to the calculation of a 'reasonable time' for completion in any given circumstances:

1. Estimates of the likely duration of construction given during negotiations.
2. Any time risk allocation in the contract.
3. The extent to which some time risks may be within the control of one or other of the parties.
4. The question of who has the burden of proof.[58]

Undue Influence

The equitable doctrine of undue influence operates to release parties from contracts that they have entered into, not as a result of improper threats, but as a result of being 'influenced' by the other party, whether intentionally or not. The precise concept may be due for reconsideration; however, at the present there are authorities that are treated as being concerned with undue influence.

'Influence' in itself is perfectly acceptable in most instances. It is only when it becomes 'undue' that the law will intervene.

Influence becomes 'undue' when an imbalance of power between the parties has been used illegitimately by the influencer. Alternatively, the word can be used simply to indicate that the level of influence is at such a level that the influenced party has lost autonomy in deciding whether to enter into a contract. This does not imply any necessary impropriety on the part of the influencer.

There is an issue of whether the concept is 'claimant-focused' or 'defendant-focused'. If it is claimant-focused, then what matters is whether the claimant acted autonomously in entering into the contract. If it is defendant-focused, then what

[57] releasing them from the liquidated damages clause.

[58] The contractor may establish that the time it has taken is reasonable or the employer must establish that it is unreasonable.

matters is whether the defendant has deliberately taken advantage of the claimant's weaker position.

The basic test in law is that it is only where there is some relationship between the parties that leads to an inequality between them that the law will intervene. The starting point for the law's analysis is therefore not the substance of the transaction, but the process by which it came about.

Generally, the claimant must prove, on the balance of probabilities, that in relation to a particular transaction, the defendant used undue influence. There is no need here for there to be a previous history of such influence. It can operate for the first time in connection with the transaction that is disputed.[59]

Value Engineering

Commonly used value techniques are defined as:

1. Value Management. This is a higher-order title and is linked to a particular project stage at which value management techniques may be applied.
2. Value Planning. Value techniques applied during the planning phases of a project.
3. Value Engineering. Value techniques applied during the design or 'engineering' phases of a project.
4. Value Analysis. Value techniques applied retrospectively to completed projects to analyse or audit the project's performance.

Value engineering is an exercise that involves most of the project team as the project develops. Its aim is to take a wider view and look at the selection of materials, plant, equipment and processes to see if a more cost-effective solution exists that will achieve the same project objectives.

In construction, value engineering involves considering the availability of materials, construction methods, transportation issues, site limitations or restrictions, planning and organisation, costs, profits and so on. Benefits that can be delivered include a reduction in life-cycle costs, improvement in quality, reduction of environmental impacts and so on.

Value engineering should start at project inception, where the benefits can be greatest; however, the contractor may also have a significant contribution to make, as long as the changes required to the contract do not affect the timescales or completion dates, or incur additional costs that outweigh the savings on offer.

Value engineering involves:

1. Identifying the main elements of a product, service or project.
2. Analysing the functions of those elements.
3. Developing alternative solutions for delivering those functions.

---

[59] *Bank of Credit and Commerce International v Aboody [1990] 1 QB 923.*

4. Assessing the alternative solutions.
5. Allocating costs to the alternative solutions.
6. Developing in more detail the alternatives with the highest likelihood of success.

Under a typical form, the contractor shall—if required by the engineer to do so, or of its own volition at any time—submit to the engineer, or require a tenderer or subcontractor to submit, a written proposal which (in the contractor's opinion) will, if adopted:

1. accelerate completion;
2. reduce the cost to the employer of executing, maintaining or operating the works;
3. improve the efficiency or value to the employer of the completed works; or
4. otherwise be of benefit to the employer.

Any proposal for value engineering shall be prepared at the cost of the contractor, and shall include the items listed in the clause (variations). In the event that the employer wishes to accept any proposal of the contractor, the engineer shall issue instructions requiring the implementation of a variation in relation thereto. If such instructions are either in the form of or subsequently confirmed in writing by way of a formal variation order or change order, the contractor shall proceed to implement said variation.

Variations and Change Orders

Changes in the work may be accomplished after execution of the contract, and without invalidating the contract, by change order.

Changes may be made by the employer in the period of performance, the design, the method or manner of performance of work, the drawings, the employer's furnished facilities, the equipment, the materials, the services or the project site, or by directing the acceleration of work or specifications of the contract (Pickavance 2010).

Changes in the work shall be performed under applicable provisions of the contract documents, and the contractor shall proceed promptly, unless otherwise provided in the change order.

A change order is a written instrument prepared and signed by the engineer stating one or more of the following:

1. A change in the period of performance, work, drawing(s) , design, method or manner of performance of the work, employer's furnished facilities, equipment, materials, services, project site or by the employer directing the acceleration of work and/or specifications.
2. The amount of the adjustment in the contract sum, if any.
3. The extent of the adjustment in the contract schedule, if any.

Change orders and variations in fact arise for many reasons other than technical problems in fulfilling the contemplated design.

The freedom to make changes after the contract has been let has become a hallmark of traditional construction practice. The engineer is given the power to order change, and this power is often exercised. Its impact on the time for completion of the works, and on their out-turn cost, can be greatly disproportionate to the extent of any particular change. The cumulative effect of such instructed changes can undermine the whole economy of a project if it is not well managed.

When ordered to carry out a variation, the contractor has the following options to choose from:

1. If the request is purportedly given under the variation clause and the contractor complies without objection, he will be estopped from subsequently claiming any right to payment other than under the terms of the contract.
2. If the request is made without any reference to the variation clause then the contractor will be entitled to a reasonable remuneration on the basis of an implied reasonable price under a separate contract.
3. On the strict basis of interpretation applied to exclusion clauses, the courts will construe restrictive provisions. For example, those requiring written instructions as a condition precedent, therefore enabling the contractor to recover a reasonable price for work outside its scope free of any such restrictions.
4. If work outside the scope is carried out, then the original pricing mechanism should only be departed from with respect to the work outside the scope and not any part of the original contract works.

Change orders and variations must also be well managed in any construction contract, and should include:

1. Clarification of time and cost risks to be borne by the principal.
2. Increased flexibility to vary the way the work is managed.
3. Detailed specification for the provision of programmes, method statements and historical progress records.
4. An ability to vary the resources, method or sequence of the works.

References

Books

Beale H (2020) Chitty on contracts, 3rd edn. Sweet & Maxwell, London
Garner BA (2019) Black's law dictionary, 11th edn. Thomson Reuters, Eagan, MN
Martinez M, Rawley D, Williams K (2012) Construction adjudication and payments handbook. OUP, Oxford
Mason J (2016) Construction law from beginner to practitioner. Routledge, Abingdon
McKendrick E (2020) Contract law: text, cases, and materials, 9th edn. OUP, Oxford

Merkin R, Saintier S (2019) Poole's casebook on contract law, 14th edn. OUP, Oxford
Pickavance K (2010) Delay and disruption in construction contracts. Sweet & Maxwell, London
Poole J (2016) Textbook on contract law, 11th edn. OUP, Oxford

Case Law

Armory v Delamirie [1722] EWHC J94
AXA Sun Life Services Plc v Campbell Martin Ltd & Ors [2011] EWCA Civ 133
Balfour Beatty Regional Construction Ltd v Grove Developments Ltd v [2016] EWCA Civ 990
Bank of Credit and Commerce International v Aboody [1990] 1 QB 923
Barton v Armstrong [1976] AC 104
Bellgrove v Eldridge [1954] 90 CLR 613
Blyth v Birmingham Waterworks Company (1856) 11 Ex Ch 781
Bramall & Ogden v Sheffield City Council (1985) 1 Con LR 30
Briginshaw v Briginshaw (1938) 60 CLR 336
Caparo Industries PLC v Dickman [1990] UKHL 2
Critchlow, J. (2007) and Hochster v. De la Tour (1853) 2 El & Bl 678; 118 ER 922
Davis Contractors v Fareham UDC [1956] AC 696
Donoghue v Stevenson (1932) AC 562
Dunlop Pneumatic Tyre Co v New Garage & Motor Co Ltd ([1915] Ac 79
Dunlop Tyre Co v Selfridge [1915] UKHL 1
Greaves v Baynham Meikle [1975] 1 WLR 1095 at 1098, CA
H W Nevill (Sunblest) v William Press [1981] 20 BLR 78
Hadley v Baxendale [1854] EWHC J70
Hedley Byrne & Co. Ltd. v Heller & Partners Ltd. [1964] AC 465
Holme v Guppy (1838) 3 M & W 387
Immingham Storage Company Ltd v Clear plc [2011] EWCA Civ 89
Interfoto Picture Library v Stiletto Visual Programmes [1987] EWCA Civ 6
James McNaghten Paper Group Ltd v Hicks Anderson & Co (1991)
JEB Fasteners Ltd v Marks, Bloom & Co (1982)
Lord Denning in Butler MachineTool v Ex-Cell-O [1977] EWCA Civ 9
Morgan Crucible v Hill Samuel Bank Ltd (1991)
Occidental Worldwide Investment Corp. v. Skibs A/S Avanti, Skibs A/S Glarona, Skibs A/S Navalis (The "Siboen" and the "Sibotre") [1976] 1 Lloyd's Rep. 293
Peak Construction (Liverpool) Ltd v McKinney Foundations Ltd. (1970) 1 BLR 111
Regional Construction Sdn. BHd. v Chung Syn Kheng Electrical Co. Bhd. [1987] 2 MLJ 763
Shanklin Pier v Detel Products [1951] 2 KB 854
Trollope & Colls Ltd v North West Metropolitan Regional Hospital [1973] 1 W.L.R. 601
Tweddle v Atkinson (1861) 1 B&S 393
Whiten v Pilot Insurance Co. 2002 S.C.R. 595.
Williams v Roffey Bros & Nicholls (Contractors) Ltd [1989] EWCA Civ 5
Yam Seng Pte Ltd v International Trade Corporation Ltd [2013] EWHC 111

Index

A. D. Haidar, *Handbook of Contract Management in Construction*,
https://doi.org/10.1007/978-3-030-72265-4

Printed in the United States
by Baker & Taylor Publisher Services